Active Prelude to Calculus

Active Prelude to Calculus

Matthew Boelkins
Grand Valley State University

Production Editor

Mitchel T. Keller
Morningside College

July 26, 2019

Cover Photo: Noah Wyn Photography

Edition: 2019

Website: http://activecalculus.org

Acknowledgements

This text began as my sabbatical project in the fall semester of 2018, during which I wrote most of the material. For the sabbatical leave, I express my deep gratitude to Grand Valley State University for its support of the project, as well as to my colleagues in the Department of Mathematics and the College of Liberal Arts and Sciences for their endorsement of the project.

The beautiful full-color .eps graphics, as well as the occasional interactive JavaScript graphics, use David Austin's Python library that employs Bill Casselman's PiScript. The .html version of the text is the result Rob Beezer's amazing work to develop the publishing language PreTeXt (formerly known as Mathbook XML); learn more at pretextbook.org. I'm grateful to the American Institute of Mathematics for hosting and funding a weeklong workshop in Mathbook XML in San Jose, CA, in April 2016, which enabled me to get started in Pre-TeXt. The ongoing support of the user group is invaluable, and David Farmer of AIM is has also been a source of major support and advocacy. Mitch Keller of Morningside College is the production editor of both Active Calculus: Single Variable and this text; his technical expertise is at.

This first public offering of the text in 2019 will benefit immensely from user feedback and suggestions. I welcome hearing from you.

Matt Boelkins
Allendale, MI
August 2019
boelkinm@gvsu.edu

Contributors

Users of the text contribute important insight: they find errors, suggest improvements, and offer feedback and impressions. I'm grateful for all of it. As you use the text, I hope you'll contact me to share anything you think could make the book better.

The following contributing editors have offered feedback that includes information about typographical errors or suggestions to improve the exposition.

MANDY FORSLUND
  GVSU

ROBERT TALBERT
  GVSU

MARCIA FROBISH
  GVSU

Our Goals

This text is designed for college students who aspire to take calculus and who either need to take a course to prepare them for calculus or want to do some additional self-study. Many of the core topics of the course will be familiar to students who have completed high school. At the same time, we take a perspective on every topic that emphasizes how it is important in calculus. This text is written in the spirit of *Active Calculus* and is especially ideal for students who will eventually study calculus from that text. The reader will find that the text requires them to engage actively with the material, to view topics from multiple perspectives, and to develop deep conceptual undersanding of ideas.

Many courses at the high school and college level with titles such as "college algebra", "pre-calculus", and "trigonometry" serve other disciplines and courses other than calculus. As such, these prerequisite classes frequently contain wide-ranging material that, while mathematically interesting and important, isn't necessary for calculus. Perhaps because of these additional topics, certain ideas that are essential in calculus are under-emphasized or ignored. In *Active Prelude to Calculus*, one of our top goals is to keep the focus narrow on the following most important ideas.

- *Functions as processes.* The mathematical concept of function is sophisticated. Understanding how a function is a special mathematical process that converts a collection of inputs to a collection of outputs is crucial for success in calculus, as calculus is the study of how functions change.

- *Average rate of change.* The central idea in differential calculus is the instantaneous rate of change of a function, which measures how fast a function's output changes with respect to changes in the input at a particular location. Because instantaneous rate of change is defined in terms of average rate of change, it's essential that students are comfortable and familiar with the idea, meaning, and applications of average rate of change.

- *Library of basic functions.* The vast majority of functions in calculus come from an algebraic combination of a collection of familiar basic functions that include power, circular, exponential, and logarithmic functions. By developing understanding of a relatively small family of basic functions and using these along with transformations to consider larger collections of functions, we work to make the central objects of calculus more intuitive and accessible.

- *Families of functions that model important phenomena.* Mathematics is the language of science, and it's remarkable how effective mathematics is at representing observable physical phenomena. From quadratic functions that model how an object falls under the influence of gravity, to shifted exponential functions that model how coffee cools, to sinusoidal functions that model how a spring-mass system oscillates, familiar basic functions find many important applications in the world around us. We regularly use these physical situations to help us see the importance of functions and to understand how families of functions that depend on different parameters are needed to represent these situations.

- *The sine and cosine are circular functions.* Many students are first introduced to the sine and cosine functions through right triangles. While this perspective is important, it is more important in calculus and other advanced courses to understand how the sine and cosine functions arise from a point traversing a circle. We take this circular function perspective early and first, and do so in order to develop deep understanding of how the familiar sine and cosine waves are generated.

- *Inverses of functions.* When a function has an inverse function, the inverse function affords us the opportunity to view an idea from a new perspective. Inverses also play a crucial role in solving algebraic equations and in determining unknown parameters in models. We emphasize the perspective that an inverse function is a process that reverses the process of the original function, as well as important basic functions that arise as inverses of other functions, especially logarithms and inverse trigonometric functions.

- *Exact values versus approximate ones.* The ability to represent numbers exactly is a powerful tool in mathematics. We regularly and consistently distinguish between a number's exact value, such as $\sqrt{2}$, and its approximation, say 1.414. This idea is also closely tied to functions and function notation: e^{-1}, $\cos(2)$, and $\ln(7)$ are all symbolic representations of exact numbers that can only be approximated by a computer.

- *Finding function formulas in applied settings.* In applied settings with unknown variables, it's especially useful to be able to represent relationships among variables, since such relationships often lead to functions whose behavior we can study. We work throughout *Active Prelude to Calculus* to ready students for problems in calculus that ask them to develop function formulas by observing relationships.

- *Long-term trends, unbounded behavior, and limits.* By working to study functions as objects themselves, we often focus on trends and overall behavior. In addition to introducing the ideas of a function being increasing or decreasing, or concave up or concave down, we also focus on using algebraic approaces to comprehend function behavior where the input and/or output increase without bound. In anticipation of calculus, we use limit notation and work to understand how this shorthand summarizes key features of functions.

Features of the Text

Instructors and students alike will find several consistent features in the presentation, including:

Motivating Questions At the start of each section, we list 2–3 *motivating questions* that provide motivation for why the following material is of interest to us. One goal of each section is to answer each of the motivating questions.

Preview Activities Each section of the text begins with a short introduction, followed by a *preview activity*. This brief reading and preview activity are designed to foreshadow the upcoming ideas in the remainder of the section; both the reading and preview activity are intended to be accessible to students *in advance* of class, and indeed to be completed by students before the particular section is to be considered in class.

Activities A typical section in the text has at least three *activities*. These are designed to engage students in an inquiry-based style that encourages them to construct solutions to key examples on their own, working in small groups or individually.

Exercises There are dozens of college algebra and trignometry texts with (collectively) tens of thousands of exercises. Rather than repeat standard and routine exercises in this text, we recommend the use of WeBWorK with its access to the Open Problem Library (OPL) and many thousands of relevant problems. In this text, each section includes a small collection of anonymous WeBWorK exercises that offer students immediate feedback without penalty, as well as 3–4 additional challenging exercises per section. Each of the non-WeBWorK exercises has multiple parts, requires the student to connect several key ideas, and expects that the student will do at least a modest amount of writing to answer the questions and explain their findings.

Graphics As much as possible, we strive to demonstrate key fundamental ideas visually, and to encourage students to do the same. Throughout the text, we use full-color[1] graphics to exemplify and magnify key ideas, and to use this graphical perspective alongside both numerical and algebraic representations of calculus.

Interactive graphics Many of the ideas of how functions behave are best understood dynamically; applets offer an often ideal format for investigations and demonstrations. *Desmos* provides a free and easy-to-use online graphing utility that we occasionally

---

[1]To keep cost low, the graphics in the print-on-demand version are in black and white. When the text itself refers to color in images, one needs to view the .html or .pdf electronically.

link to and often direct students to use. Thanks to David Austin, there are also select interactive javascript figures within the text itself.

Summary of Key Ideas Each section concludes with a summary of the key ideas encountered in the preceding section; this summary normally reflects responses to the motivating questions that began the section.

Students! Read this!

This book is different.

The text is available in three different formats: HTML, PDF, and print, each of which is available via links on the landing page at https://activecalculus.org/. The first two formats are free. If you are going to use the book electronically, the best mode is the HTML. The HTML version looks great in any browser, including on a smartphone, and the links are much easier to navigate in HTML than in PDF. Some particular direct suggestions about using the HTML follow among the next few paragraphs; alternatively, you can watch this short video from the author (based on using the text *Active Calculus*, which is similar). It is also wise to download and save the PDF, since you can use the PDF offline, while the HTML version requires an internet connection. An inexpensive print copy is available on Amazon.

This book is intended to be read sequentially and engaged with, much more than to be used as a lookup reference. For example, each section begins with a short introduction and a Preview Activity; you should read the short introduction and complete the Preview Activity prior to class. Your instructor may require you to do this. Most Preview Activities can be completed in 15-20 minutes and are intended to be accessible based on the understanding you have from preceding sections.

As you use the book, think of it as a workbook, not a worked-book. There is a great deal of scholarship that shows people learn better when they interactively engage and struggle with ideas themselves, rather than passively watch others. Thus, instead of reading worked examples or watching an instructor complete examples, you will engage with Activities that prompt you to grapple with concepts and develop deep understanding. You should expect to spend time in class working with peers on Activities and getting feedback from them and from your instructor. You can ask your instructor for a copy of the PDF file that has only the activities along with room to record your work. Your goal should be to do all of the activities in the relevant sections of the text and keep a careful record of your work.

Each section concludes with a Summary. Reading the Summary after you have read the section and worked the Activities is a good way to find a short list of key ideas that are most essential to take from the section. A good study habit is to write similar summaries in your own words.

At the end of each section, you'll find two types of Exercises. First, there are several anonymous WeBWorK exercises. These are online, interactive exercises that allow you to submit answers for immediate feedback with unlimited attempts without penalty; to submit answers, you have to be using the HTML version of the text (see this short video on the HTML version that includes a WeBWorK demonstration). You should use these exercises as a way to test your understanding of basic ideas in the preceding section. If your institution uses WeBWorK, you may also need to log in to a server as directed by your instructor to complete assigned WeBWorK sets as part of your course grade. The WeBWorK exercises included in this text are ungraded and not connected to any individual account. Following the WeBWorK exercises there are 3-4 additional challenging exercises that are designed to encourage you to connect ideas, investigate new situations, and write about your understanding.

The best way to be successful in mathematics generally and calculus specifically is to strive to make sense of the main ideas. We make sense of ideas by asking questions, interacting with others, attempting to solve problems, making mistakes, revising attempts, and writing and speaking about our understanding. This text has been designed to help you make sense of key ideas that are needed in calculus and to help you be well-prepared for success in calculus; we wish you the very best as you undertake the large and challenging task of doing so.

Instructors! Read this!

This book is different. Before you read further, first read "Students! Read this!" as well as "Our Goals".

Among the three formats (HTML, PDF, print), the HTML is optimal for display in class if you have a suitable projector. The HTML is also best for navigation, as links to internal and external references are much more obvious. We recommend saving a downloaded version of the PDF format as a backup in the event you don't have internet access. It's a good idea for each student to have a printed version of the Activities Workbook, which you can acquire as a PDF document by direct request to the author (boelkinm at gvsu dot edu); many instructors use the PDF to have coursepacks printed for students to purchase from their local bookstore.

The text is written so that, on average, one section corresponds to two hours of class meeting time. A typical instructional sequence when starting a new section might look like the following:

- Students complete a Preview Activity in advance of class. Class begins with a short debrief among peers followed by all class discussion. (5-10 minutes)

- Brief lecture and discussion to build on the preview activity and set the stage for the next activity. (5-10 minutes)

- Students engage with peers to work on and discuss the first activity in the section. (15-20 minutes)

- Brief discussion and possibly lecture to reach closure on the preceding activity, followed by transition to new ideas. (Varies, but 5-15 minutes)

- Possibly begin next activity.

The next hour of class would be similar, but without the Preview Activity to complete prior to class: the principal focus of class will be completing 2 activities. Then rinse and repeat.

We recommend that instructors use appropriate incentives to encourage students to complete Preview Activities prior to class. Having these be part of completion-based assignments that count 5% of the semester grade usually results in the vast majority of students completing the vast majority of the previews. If you'd like to see a sample syllabus for how to organize a course and weight various assignments, you can request one via email to the author.

Note that the WeBWorK exercises in the HTML version are anonymous and there's not a way to track students' engagement with them. These are intended to be formative for students and provide them with immediate feedback without penalty. If your institution is a WeBWorK user, in the near future we will have sets of .def files that correspond to the sections in the text; these will be available upon request to the author.

The PreTeXt source code for the text can be found on GitHub. If you find errors in the text or have other suggestions, you can file an issue on GitHub or email the author directly. To engage with instructors who use the text, we maintain both an email list and the Open Calculus blog; you can request that your address be added to the email list by contacting the author. Finally, if you're interested in a video presentation on using the similar *Active Calculus* text, you can see this online video presentation to the MIT Electronic Seminar on Mathematics Education; at about the 17-minute mark, the portion begins where we demonstrate features of and how to use the text.

Thank you for considering *Active Prelude to Calculus* as a resource to help your students develop deep understanding of the subject. I wish you the very best in your work and hope to hear from you.

Contents

CHAPTER 1

Relating Changing Quantities

1.1 Changing in Tandem

Motivating Questions

- If we have two quantities that are changing in tandem, how can we connect the quantities and understand how change in one affects the other?

- When the amount of water in a tank is changing, what behaviors can we observe?

Mathematics is the art of making sense of patterns. One way that patterns arise is when two quantities are changing in tandem. In this setting, we may make sense of the situation by expressing the relationship between the changing quantities through words, through images, through data, or through a formula.

Preview Activity 1.1.1. Suppose that a rectangular aquarium is being filled with water. The tank is 4 feet long by 2 feet wide by 3 feet high, and the hose that is filling the tank is delivering water at a rate of 0.5 cubic feet per minute.

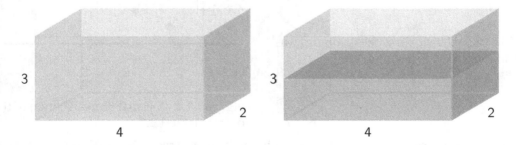

Figure 1.1.1: The empty aquarium.   Figure 1.1.2: The aquarium, partially filled.

a. What are some different quantities that are changing in this scenario?

b. After 1 minute has elapsed, how much water is in the tank? At this moment,

> how deep is the water?
>
> c. How much water is in the tank and how deep is the water after 2 minutes? After 3 minutes?
>
> d. How long will it take for the tank to be completely full? Why?

1.1.1 Using Graphs to Represent Relationships

In Preview Activity 1.1.1, we saw how several changing quantities were related in the setting of an aquarium filling with water: time, the depth of the water, and the total amount of water in the tank are all changing, and any pair of these quantities changes in related ways. One way that we can make sense of the situation is to record some data in a table. For instance, observing that the tank is filling at a rate of 0.5 cubic feet per minute, this tells us that after 1 minute there will be 0.5 cubic feet of water in the tank, and after 2 minutes there will be 1 cubic foot of water in the tank, and so on. If we let t denote the time in minutes and V the amount of water in the tank at time t, we can represent the relationship between these quantities through Table 1.1.3.

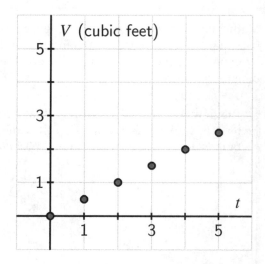

t	V
0	0.0
1	0.5
2	1.0
3	1.5
4	2.0
5	2.5

Table 1.1.3: Data for how the volume of water in the tank changes with time.

Figure 1.1.4: A visual representation of the data in Table 1.1.3.

We can also represent this data in a graph by plotting ordered pairs (t, V) on a system of coordinate axes, where t represents the horizontal distance of the point from the origin, $(0, 0)$, and V represents the vertical distance from $(0, 0)$. The visual representation of the table of values from Table 1.1.3 is seen in the graph in Figure 1.1.4.

Sometimes it is possible to use variables and one or more equations to connect quantities that are changing in tandem. In the aquarium example from the preview activity, we can observe that the volume, V, of a rectangular box that has length l, width w, and height h is given by

$$V = l \cdot w \cdot h,$$

and thus, since the water in the tank will always have length $l = 4$ feet and width $w = 2$ feet, the volume of water in the tank is directly related to the depth of water in the tank by the equation

$$V = 4 \cdot 2 \cdot h = 8h.$$

Depending on which variable we solve for, we can either see how V depends on h through the equation $V = 8h$, or how h depends on V via the equation $h = \frac{1}{8}V$. From either perspective, we observe that as depth or volume increases, so must volume or depth correspondingly increase.

Activity 1.1.2. Consider a tank in the shape of an inverted circular cone (point down) where the tank's radius is 2 feet and its depth is 4 feet. Suppose that the tank is being filled with water that is entering at a constant rate of 0.75 cubic feet per minute.

a. Sketch a labeled picture of the tank, including a snapshot of there being water in the tank prior to the tank being completely full.

b. What are some quantities that are changing in this scenario? What are some quantities that are not changing?

c. Fill in the following table of values to determine how much water, V, is in the tank at a given time in minutes, t, and thus generate a graph of the relationship between volume and time by plotting the data on the provided axes.

t	V
0	
1	
2	
3	
4	
5	

Table 1.1.5: Table to record data on volume and time in the conical tank.

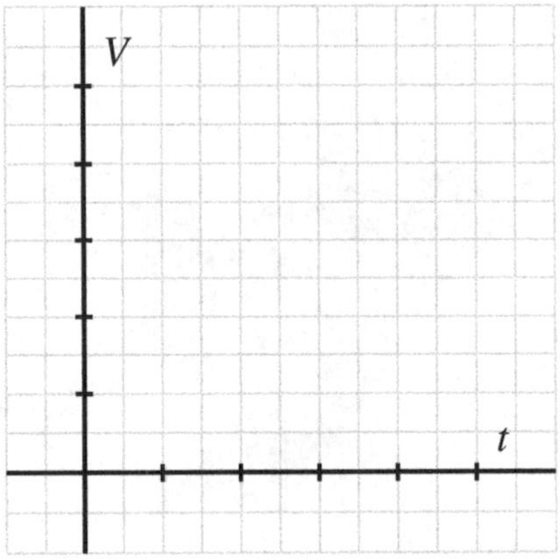

Figure 1.1.6: How volume and time change in tandem in the conical tank.

d. Finally, think about how the height of the water changes in tandem with time. Without attempting to determine specific values of h at particular values of t, how would you expect the data for the relationship between h and t to appear? Use the provided axes to sketch at least two possibilities; write at least one sentence to explain how you think the graph should appear.

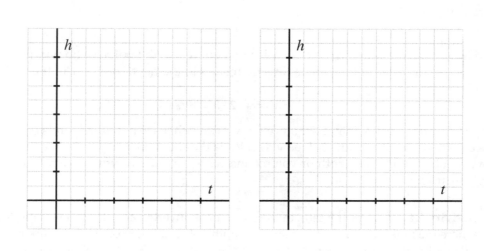

1.1.2 Using Algebra to Add Perspective

One of the ways that we make sense of mathematical ideas is to view them from multiple perspectives. We may use different means to establish different points of view: words, numerical data, graphs, or symbols. In addition, sometimes by changing our perspective within a particular approach we gain deeper insight.

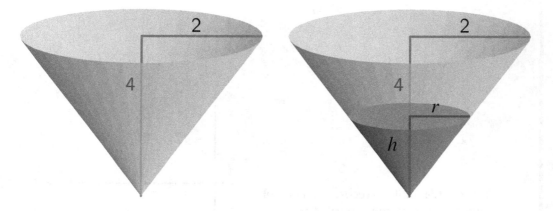

Figure 1.1.7: The empty conical tank.   **Figure 1.1.8:** The conical tank, partially filled.

If we consider the conical tank discussed in Activity 1.1.2, as seen in Figure 1.1.7 and Figure 1.1.8, we can use algebra to better understand some of the relationships among changing quantities. The volume of a cone with radius r and height h is given by the formula

$$V = \frac{1}{3}\pi r^2 h.$$

Note that at any time while the tank is being filled, r (the radius of the surface of the water), h (the depth of the water), and V (the volume of the water) are all changing; moreover, all are connected to one another. Because of the constraints of the tank itself (with radius 2 feet and depth 4 feet), it follows that as the radius and height of the water change, they always do so in the proportion

$$\frac{r}{h} = \frac{2}{4}.$$

Solving this last equation for r, we see that $r = \frac{1}{2}h$; substituting this most recent result in the equation for volume, it follows that

$$V = \frac{1}{3}\pi \left(\frac{1}{2}h\right)^2 h = \frac{\pi}{12}h^3.$$

This most recent equation helps us understand how V and h change in tandem. We know from our earlier work that the volume of water in the tank increases at a constant rate of 0.75 cubic feet per minute. This leads to the data shown in Table 1.1.9.

t	0	1	2	3	4	5
V	0.0	0.75	1.5	2.25	3.0	3.75

Table 1.1.9: How time and volume change in tandem in a conical tank.

With the equation $V = \frac{\pi}{12}h^3$, we can now also see how the height of the water changes in tandem with time. Solving the equation for h, note that $h^3 = \frac{12}{\pi}V$, and therefore

$$h = \sqrt[3]{\frac{12}{\pi}V}. \tag{1.1.1}$$

Thus, when $V = 0.75$, it follows that $h = \sqrt[3]{\frac{12}{\pi}0.75} \approx 1.42$. Executing similar computations with the other values of V in Table 1.1.9, we get the following updated data that now includes h.

t	0	1	2	3	4	5
V	0.0	0.75	1.5	2.25	3.0	3.75
h	0.0	1.42	1.79	2.05	2.25	2.43

Table 1.1.10: How time, volume, and height change in concert in a conical tank.

Plotting this data on two different sets of axes, we see the different ways that V and h change with t. Whereas volume increases at a constant rate, as seen by the straight line appearance of the points in Figure 1.1.11, we observe that the water's height increases in a way that it rises more slowly as time goes on, as shown by the way the curve the points lie on in Figure 1.1.12 "bends down" as time passes.

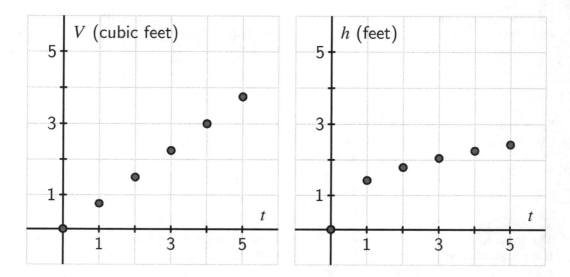

Figure 1.1.11: Plotting V versus t.   **Figure 1.1.12:** Plotting h versus t.

These different behaviors make sense because of the shape of the tank. Since at first there is less volume relative to depth near the cone's point, as water flows in at a constant rate, the water's height will rise quickly. But as time goes on and more water is added at the same rate, there is more space for the water to fill in order to make the water level rise, and thus the water's heigh rises more and more slowly as time passes.

Activity 1.1.3. Consider a tank in the shape of a sphere where the tank's radius is 3 feet. Suppose that the tank is initially completely full and that it is being drained by a pump at a constant rate of 1.2 cubic feet per minute.

   a. Sketch a labeled picture of the tank, including a snapshot of some water remaining in the tank prior to the tank being completely empty.

   b. What are some quantities that are changing in this scenario? What are some quantities that are not changing?

   c. Recall that the volume of a sphere of radius r is $V = \frac{4}{3}\pi r^3$. When the tank is completely full at time $t = 0$ right before it starts being drained, how much water is present?

   d. How long will it take for the tank to drain completely?

   e. Fill in the following table of values to determine how much water, V, is in the tank at a given time in minutes, t, and thus generate a graph of the relationship between volume and time. Write a sentence to explain why the data's graph appears the way that it does.

t	V
0	
20	
40	
60	
80	
94.24	

Table 1.1.13: Data for how volume and time change together.

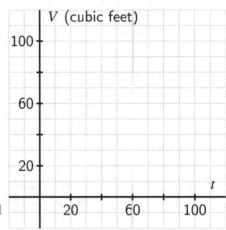

Figure 1.1.14: A plot of how volume and time change in tandem in a draining spherical tank.

f. Finally, think about how the height of the water changes in tandem with time. What is the height of the water when $t = 0$? What is the height when the tank is empty? How would you expect the data for the relationship between h and t to appear? Use the provided axes to sketch at least two possibilities; write at least one sentence to explain how you think the graph should appear.

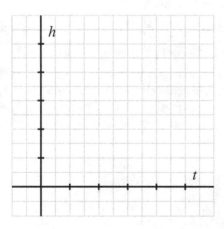
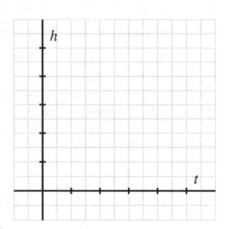

1.1.3 Summary

- When two related quantities are changing in tandem, we can better understand how change in one affects the other by using data, graphs, words, or algebraic symbols to express the relationship between them. See, for instance, Table 1.1.9, Figure 1.1.11, 1.1.12, and Equation (1.1.1) that together help explain how the height and volume of water in a conical tank change in tandem as time changes.

- When the amount of water in a tank is changing, we can observe other quantities that change, depending on the shape of the tank. For instance, if the tank is conical, we can consider both the changing height of the water and the changing radius of the surface of the water. In addition, whenever we think about a quantity that is changing as time passes, we note that time itself is changing.

1.1.4 Exercises

1. The graph below shows the fuel consumption (in miles per gallon, mpg) of a car driving at various speeds (in miles per hour, mph).

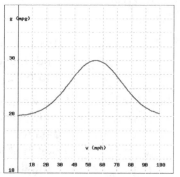

(a) How much gas is used on a 400 mile trip at 80 mph?

(b) How much gas is saved by traveling 60 mph instead of 70 mph on a 600 mile trip?

(c) According to this graph, what is the most fuel efficient speed to travel?

2. Suppose we have an unusual tank whose base is a perfect sphere with radius 3 feet, and then atop the spherical base is a cylindrical "chimney" that is a circular cylinder of radius 1 foot and height 2 feet, as shown in Figure 1.1.15. The tank is initially empty, but then a spigot is turned on that pumps water into the tank at a constant rate of 1.25 cubic feet per minute.

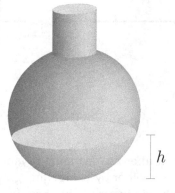

Figure 1.1.15: A spherical tank with a cylindrical chimney.

Let V denote the total volume of water (in cubic feet) in the tank at any time t (in minutes), and h the depth of the water (in feet) at time t.

   a. It is possible to use calculus to show that the total volume this tank can hold is $V_{full} = \pi(22 + \frac{38}{3}\sqrt{2}) \approx 119.12$ cubic feet. In addition, the actual height of the tank (from the bottom of the spherical base to the top of the chimney) is $h_{full} = \sqrt{8}+5 \approx 7.83$ feet. How long does it take the tank to fill? Why?

   b. On the blank axes provided below, sketch (by hand) possible graphs of how V and t change in tandem and how h and t change in tandem.

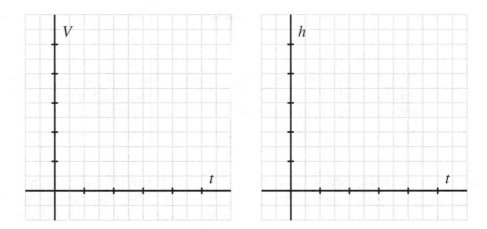

For each graph, label any ordered pairs on the graph that you know for certain, and write at least one sentence that explains why your graphs have the shape they do.

   c. How would your graph(s) change (if at all) if the chimney was shaped like an inverted cone instead of a cylinder? Explain and discuss.

3.   Suppose we have a tank that is a perfect sphere with radius 6 feet. The tank is initially empty, but then a spigot is turned on that is pumping water into the tank in a very special way: the faucet is regulated so that the depth of water in the tank is increasing at a constant rate of 0.4 feet per minute.

Let V denote the total volume of water (in cubic feet) in the tank at any time t (in minutes), and h the depth of the water (in feet) at given time t.

   a. How long does it take the tank to fill? What will the values of V and h be at the moment the tank is full? Why?

   b. On the blank axes provided below, sketch (by hand) possible graphs of how V and t change in tandem and how h and t change in tandem.

For each graph, label any ordered pairs on the graph that you know for certain, and write at least one sentence that explains why your graphs have the shape they do.

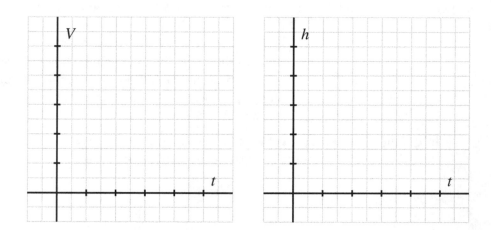

c. How do your responses change if the tank stays the same but instead the tank is initially full and the tank drains in such a way that the height of the water is always decreasing at a constant rate of 0.25 feet per minute?

4. The relationship between the position, s, of a car driving on a straight road at time t is given by the graph pictured at left in Figure 1.1.16. The car's position[1] has units measured in thousands of feet while time is measured in minutes. For instance, the point $(4, 6)$ on the graph indicates that after 4 minutes, the car has traveled 6000 feet from its starting location.

   a. Write several sentences that explain the how the car is being driven and how you make these conclusions from the graph.

   b. How far did the car travel between $t = 2$ and $t = 10$?

   c. Does the car ever travel in reverse? Why or why not? If not, how would the graph have to look to indicate such motion?

   d. On the blank axes in Figure 1.1.16, plot points or sketch a curve to describe the behavior of a car that is driven in the following way: from $t = 0$ to $t = 5$ the car travels straight down the road at a constant rate of 1000 feet per minute. At $t = 5$, the car pulls over and parks for 2 full minutes. Then, at $t = 7$, the car does an abrupt U-turn and returns in the opposite direction at a constant rate of 800 feet per minute for 5 additional minutes. As part of your work, determine (and label) the car's location at several additional points in time beyond $t = 0, 5, 7, 12$.

---

[1]You can think of the car's position like mile-markers on a highway. Saying that $s = 500$ means that the car is located 500 feet from "marker zero" on the road.

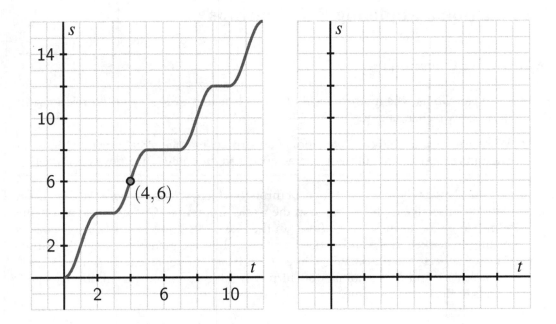

Figure 1.1.16: A graph of the relationship betwen a car's position *s* and time *t*

1.2 Functions: Modeling Relationships

Motivating Questions

- How can we use the mathematical idea of a function to represent the relationship between two changing quantities?

- What are some formal characteristics of an abstract mathematical function? how do we think differently about these characteristics in the context of a physical model?

A mathematical model is an abstract concept through which we use mathematical language and notation to describe a phenomenon in the world around us. One example of a mathematical model is found in *Dolbear's Law*[1]. In the late 1800s, the physicist Amos Dolbear was listening to crickets chirp and noticed a pattern: how frequently the crickets chirped seemed to be connected to the outside temperature. If we let T represent the temperature in degrees Fahrenheit and N the number of chirps per minute, we can summarize Dolbear's observations in the following table.

N (chirps per minute)	40	80	120	160
T (° Fahrenheit)	50°	60°	70°	80°

Table 1.2.1: Data for Dolbear's observations.

For a mathematical model, we often seek an algebraic formula that captures observed behavior accurately and can be used to predict behavior not yet observed. For the data in Table 1.2.1, we observe that each of the ordered pairs in the table make the equation

$$T = 40 + 0.25N \tag{1.2.1}$$

true. For instance, $70 = 40 + 0.25(120)$. Indeed, scientists who made many additional cricket chirp observations following Dolbear's initial counts found that the formula in Equation (1.2.1) holds with remarkable accuracy for the snowy tree cricket in temperatures ranging from about 50° F to 85° F.

> **Preview Activity 1.2.1.** Use Equation (1.2.1) to respond to the questions below.
>
> a. If we hear snowy tree crickets chirping at a rate of 92 chirps per minute, what does Dolbear's model suggest should be the outside temperature?
>
> b. If the outside temperature is 77° F, how many chirps per minute should we expect to hear?
>
> c. Is the model valid for determining the number of chirps one should hear when the outside temperature is 35° F? Why or why not?

---

[1]You can read more in the Wikipedia entry for Dolbear's Law, which has proven to be remarkably accurate for the behavior of snowy tree crickets. For even more of the story, including a reference to this phenomenon on the popular show *The Big Bang Theory*, see this article.

    d. Suppose that in the morning an observer hears 65 chirps per minute, and several hours later hears 75 chirps per minute. How much has the temperature risen between observations?

    e. Dolbear's Law is known to be accurate for temperatures from 50° to 85°. What is the fewest number of chirps per minute an observer could expect to hear? the greatest number of chirps per minute?

1.2.1 Functions

The mathematical concept of a *function* is one of the most central ideas in all of mathematics, in part since functions provide an important tool for representing and explaining patterns. At its core, a function is a repeatable process that takes a collection of input values and generates a corresponding collection of output values with the property that if we use a particular single input, the process always produces exactly the same single output.

For instance, Dolbear's Law in Equation (1.2.1) provides a process that takes a given number of chirps between 40 and 180 per minute and reliably produces the corresponding temperature that corresponds to the number of chirps, and thus this equation generates a function. We often give functions shorthand names; using "D" for the "Dolbear" function, we can represent the process of taking inputs (observed chirp rates) to outputs (corresponding temperatures) using arrows:

$$80 \xrightarrow{D} 60$$

$$120 \xrightarrow{D} 70$$

$$N \xrightarrow{D} 40 + 0.25N$$

Alternatively, for the relationship "$80 \xrightarrow{D} 60$" we can also use the equivalent notation "$D(80) = 60$" to indicate that Dolbear's Law takes an input of 80 chirps per minute and produces a corresponding output of 60 degrees Fahrenheit. More generally, we write "$T = D(N) = 40 + 0.25N$" to indicate that a certain temperature, T, is determined by a given number of chirps per minute, N, according to the process $D(N) = 40 + 0.25N$.

Tables and graphs are particularly valuable ways to characterize and represent functions. For the current example, we summarize some of the data the Dolbear function generates in Table 1.2.2 and plot that data along with the underlying curve in Figure 1.2.3.

When a point such as $(120, 70)$ in Figure 1.2.3 lies on a function's graph, this indicates the correspondence between input and output: when the value 120 chirps per minute is entered in the function D, the result is 70 degrees Fahrenheit. More concisely, $D(120) = 70$. Aloud, we read "D of 120 is 70".

For most important concepts in mathematics, the mathematical community decides on formal definitions to ensure that we have a shared language of understanding. In this text, we will use the following definition of the term "function".

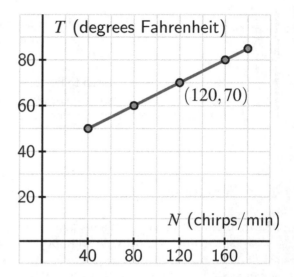

N	T
40	50
80	60
120	70
160	80
180	85

Table 1.2.2: Data for the function $T = D(N) = 40 + 0.25N$.

Figure 1.2.3: Graph of data from the function $T = D(N) = 40 + 0.25N$ and the underlying curve.

Definition 1.2.4 A **function** is a process that may be applied to a collection of input values to produce a corresponding collection of output values in such a way that the process produces one and only one output value for any single input value. ◊

If we name a given function F and call the collection of possible inputs to F the set A and the corresponding collection of potential outputs B, we say "F is a function from A to B," and sometimes write "$F : A \rightarrow B$." When a particular input value to F, say t, produces a corresponding output z, we write "$F(t) = z$" and read this symbolic notation as "F of t is z." We often call t the *independent variable* and z the *dependent variable* , since z is a function of t.

Definition 1.2.5 Let F be a function from A to B. The set A of possible inputs to F is called the **domain** of F; the set B of potential outputs from F is called the **codomain** of F. ◊

For the Dolbear function $D(N) = 40 + 0.25N$ in the context of modeling temperature as a function of the number of cricket chirps per minute, the domain of the function is $A = [40, 180]^2$ and the codomain is "all Fahrenheit temperatures". The codomain of a function is the collection of *possible* outputs, which we distinguish from the collection of *actual* ouputs.

Definition 1.2.6 Let F be a function from A to B. The **range** of F is the collection of all actual outputs of the function. That is, the range is the collection of all elements y in B for which it is possible to find an element x in A such that $F(x) = y$. ◊

In many situations, the range of a function is much more challenging to determine than its codomain. For the Dolbear function, the range is straightforward to find by using the graph shown in Figure 1.2.3: since the actual outputs of D fall between $T = 50$ and $T = 85$ and

---

²The notation "$[40, 180]$" means "the collection of all real numbers x that satisfy $40 \le x \le 80$" and is sometimes called "interval notation".

include every value in that interval, the range of D is $[50, 80]$.

The range of any function is always a subset of the codomain. It is possible for the range to equal the codomain.

Activity 1.2.2. Consider a spherical tank of radius 4 m that is filling with water. Let V be the volume of water in the tank (in cubic meters) at a given time, and h the depth of the water (in meters) at the same time. It can be shown using calculus that V is a function of h according to the rule

$$V = f(h) = \frac{\pi}{3}h^2(12 - h).$$

a. What values of h make sense to consider in the context of this function? What values of V make sense in the same context?

b. What is the domain of the function f in the context of the spherical tank? Why? What is the corresponding codomain? Why?

c. Determine and interpret (with appropriate units) the values $f(2)$, $f(4)$, and $f(8)$. What is important about the value of $f(8)$?

d. Consider the claim: "since $f(9) = \frac{\pi}{3}9^2(12 - 9) = 81\pi \approx 254.47$, when the water is 9 meters deep, there is about 254.47 cubic meters of water in the tank". Is this claim valid? Why or why not? Further, does it make sense to observe that "$f(13) = -\frac{169\pi}{3}$"? Why or why not?

e. Can you determine a value of h for which $f(h) = 300$ cubic meters?

1.2.2 Comparing models and abstract functions

Again, a mathematical model is an abstract concept through which we use mathematical language and notation to describe a phenomenon in the world around us. So far, we have considered two different examples: the Dolbear function, $T = D(N) = 40 + 0.25N$, that models how Fahrenheit temperature is a function of the number of cricket chirps per minute and the function $V = f(h) = \frac{\pi}{3}h^2(12-h)$ that models how the volume of water in a spherical tank of radius 4 m is a function of the depth of the water in the tank. While often we consider a function in the physical setting of some model, there are also many occasions where we consider an abstract function for its own sake in order to study and understand it.

Example 1.2.7 A parabola and a falling ball. Calculus shows that for a tennis ball tossed vertically from a window 48 feet above the ground at an initial vertical velocity of 32 feet per second, the ball's height above the ground at time t (where $t = 0$ is the instant the ball is tossed) can be modeled by the function $h = g(t) = -16t^2 + 32t + 48$. Discuss the differences between the model g and the abstract function f determined by $y = f(x) = -16x^2 + 32x + 48$.

Solution. We start with the abstract function $y = f(x) = -16x^2 + 32x + 48$. Absent a physical context, we can investigate the behavior of this function by computing function values, plotting points, and thinking about its overall behavior. We recognize the function

f as quadratic[3], noting that it opens down because of the leading coefficient of -16, with vertex located at $x = \frac{-32}{2(-16)} = 1$, y-intercept at $(0, 48)$, and with x-intercepts at $(-1, 0)$ and $(3, 0)$ because

$$-16x^2 + 32x + 48 = -16(x^2 - 2x - 3) = -16(x - 3)(x + 1).$$

Computing some additional points to gain more information, we see both the data in Table 1.2.8 and the corresponding graph in Figure 1.2.9.

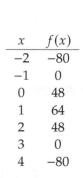

x	$f(x)$
-2	-80
-1	0
0	48
1	64
2	48
3	0
4	-80

Table 1.2.8: Data for the function $y = f(x) = -16x^2 + 32x + 48$.

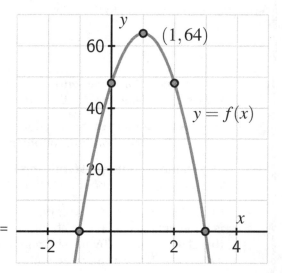

Figure 1.2.9: Graph of the function $y = f(x)$ and some data from the table.

For this abstract function, its domain is "all real numbers" since we may input any real number x we wish into the formula $f(x) = -16x^2 + 32x + 48$ and have the result be defined. Moreover, taking a real number x and processing it in the formula $f(x) = -16x^2 + 32x + 48$ will produce another real number. This tells us that the codomain of the abstract function f is also "all real numbers." Finally, from the graph and the data, we observe that the largest possible output of the function f is $y = 64$. It is apparent that we can generate any y-value less than or equal to 64, and thus the range of the abstract function f is all real numbers less than or equal to 64. We denote this collection of real numbers using the shorthand interval notation $(-\infty, 64]$.[4]

Next, we turn our attention to the model $h = g(t) = -16t^2 + 32t + 48$ that represents the height of the ball, h, in feet t seconds after the ball in initially launched. Here, the big difference is the domain, codomain, and range associated with the model. Since the model takes effect once the ball is tossed, it only makes sense to consider the model for input values $t \geq 0$. Moreover, because the model ceases to apply once the ball lands, it is only valid for $t \leq 3$. Thus, the domain of g is $[0, 3]$. For the codomain, it only makes sense to consider values of h that are nonnegative. That is, as we think of *potential* outputs for the model, then can only be in the interval $[0, \infty)$. Finally, we can consider the graph of the model on the given domain in Figure 1.2.11 and see that the range of the model is $[0, 64]$, the collection of all heights between its lowest (ground level) and its largest (at the vertex).

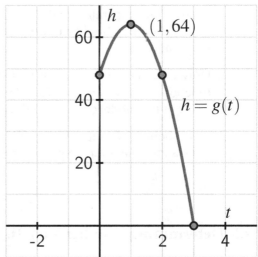

t	$g(t)$
0	48
1	64
2	48
3	0

Table 1.2.10: Data for the model $h = g(t) = -16t^2 + 32t + 48$.

Figure 1.2.11: Graph of the model $h = g(t)$ and some data from the table.

□

Activity 1.2.3. Consider a spherical tank of radius 4 m that is completely full of water. Suppose that the tank is being drained by regulating an exit valve in such a way that the height of the water in the tank is always decreasing at a rate of 0.5 meters per minute. Let V be the volume of water in the tank (in cubic meters) at a given time t (in minutes), and h the depth of the water (in meters) at the same time. It can be shown using calculus that V is a function of t according to the model

$$V = p(t) = \frac{256\pi}{3} - \frac{\pi}{24}t^2(24 - t).$$

In addition, let $h = q(t)$ be the function whose output is the depth of the water in the tank at time t.

a. What is the height of the water when $t = 0$? When $t = 1$? When $t = 2$? How long will it take the tank to completely drain? Why?

b. What is the domain of the model $h = q(t)$? What is the domain of the model $V = p(t)$?

c. How much water is in the tank when the tank is full? What is the range of the model $h = q(t)$? What is the range of the model $V = p(t)$?

d. We will frequently use a graphing utility to help us understand function behavior, and strongly recommend *Desmos* because it is intuitive, online, and free.[5]

   In this prepared *Desmos* worksheet, you can see how we enter the (abstract) function $V = p(t) = \frac{256\pi}{3} - \frac{\pi}{24}t^2(24 - t)$, as well as the corresponding graph

---

[3]We will engage in a brief review of quadratic functions in Section 1.5

[4]The notation $(-infty, 64]$ stands for all the real numbers that lie to the left of an including 64. The "$-\infty$" indicates that there is no left-hand bound on the interval.

the program generates. Make as many observations as you can about the model $V = p(t)$. You should discuss its shape and overall behavior, its domain, its range, and more.

e. How does the model $V = p(t) = \frac{256\pi}{3} - \frac{\pi}{24}t^2(24 - t)$ differ from the abstract function $y = r(x) = \frac{256\pi}{3} - \frac{\pi}{24}x^2(24 - x)$? In particular, how do the domain and range of the model differ from those of the abstract function, if at all?

f. How should the graph of the height function $h = q(t)$ appear? Can you determine a formula for q? Explain your thinking.

1.2.3 Determining whether a relationship is a function or not

To this point in our discussion of functions, we have mostly focused on what the function process may model and what the domain, codomain, and range of a model or abstract function are. It is also important to take note of another part of Definition 1.2.4: "... the process produces one and only one output value for any single input value". Said differently, if a relationship or process ever associates a single input with two or more different outputs, the process cannot be a function.

Example 1.2.12 Is the relationship between people and phone numbers a function?

Solution. No, this relationship is not a function. A given individual person can be associated with more than one phone number, such as their cell phone and their work telephone. This means that we can't view phone numbers as a function of people: one input (a person) can lead to two different outputs (phone numbers). We also can't view people as a function of phone numbers, since more than one person can be associated with a phone number, such as when a family shares a single phone at home. □

Example 1.2.13 The relationship between x and y that is given in the following table where we attempt to view y as depending on x.

x	1	2	3	4	5
y	13	11	10	11	13

Table 1.2.14: A table that relates x and y values.

Solution. The relationship between y and x in Table 1.2.14 allows us to think of y as a function of x since each particular input is associated with one and only one output. If we name the function f, we can say for instance that $f(4) = 11$. Moreover, the domain of f is the set of inputs $\{1, 2, 3, 4, 5\}$, and the codomain (which is also the range) is the set of outputs $\{10, 11, 13\}$. □

---

[5]To learn more about *Desmos*, see their outstanding online tutorials.

Activity 1.2.4. Each of the following prompts describes a relationship between two quantities. For each, your task is to decide whether or not the relationship can be thought of as a function. If not, explain why. If so, state the domain and codomain of the function and write at least one sentence to explain the process that leads from the collection of inputs to the collection of outputs.

a. The relationship between x and y in each of the graphs below (address each graph separately as a potential situation where y is a function of x). In Figure 1.2.15, any point on the circle relates x and y. For instance, the y-value $\sqrt{7}$ is related to the x-value -3. In Figure 1.2.16, any point on the blue curve relates x and y. For instance, when $x = -1$, the corresponding y-value is $y = 3$. An unfilled circle indicates that there is not a point on the graph at that specific location.

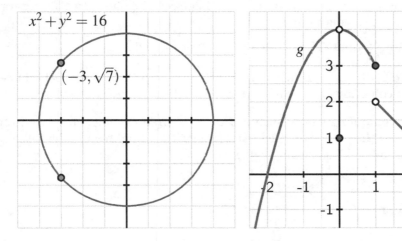

Figure 1.2.15: A circle of radius 4 centered at $(0, 0)$.

Figure 1.2.16: A graph of a possible function g.

b. The relationship between the day of the year and the value of the S&P500 stock index (at the close of trading on a given day), where we attempt to consider the index's value (at the close of trading) as a function of the day of the year.

c. The relationship between a car's velocity and its odometer, where we attempt to view the car's odometer reading as a function of its velocity.

d. The relationship between x and y that is given in the following table where we attempt to view y as depending on x.

x	1	2	3	2	1
y	11	12	13	14	15

Table 1.2.17: A table that relates x and y values.

For a relationship or process to be a function, each individual input must be associated with one and only one output. Thus, the usual way that we demonstrate a relationship or process is not a function is to find a particular input that is associated with two or more outputs. When the relationship is given graphically, such as in Figure 1.2.15, we can use the vertical line test to determine whether or not the graph represents a function.

Vertical Line Test.

A graph in the plane represents a function if and only if every vertical line intersects the graph at most once. When the graph passes this test, the vertical coordinate of each point on the graph can be viewed as a function of the horizontal coordinate of the point.

Since the vertical line $x = -3$ passes through the circle in Figure 1.2.15 at both $y = -\sqrt{7}$ and $y = \sqrt{7}$, the circle does not represent a relationship where y is a function of x. However, since any vertical line we draw in Figure 1.2.16 intersects the blue curve at most one time, the graph indeed represents a function.

We conclude with a formal definition of the graph of a function.

Definition 1.2.18 Let $F : A \rightarrow B$, where A and B are each collections of real numbers. The **graph** of F is the collection of all ordered pairs (x, y) that satisfy $y = F(x)$. ◊

When we use a computing device such as *Desmos* to graph a function g, the program is generating a large collection of ordered pairs $(x, g(x))$, plotting them in the x-y plane, and connecting the points with short line segments.

1.2.4 Summary

- A function is a process that generates a relationship between two collections of quantities. The function associates each member of a collection of input values with one and only one member of the collection of output values. A function can be described or defined by words, by a table of values, by a graph, or by a formula.

- Functions may be viewed as mathematical objects worthy of study for their own sake and also as models that represent physical phenomena in the world around us. Every function or model has a domain (the set of possible or allowable input values), a codomain (the set of possible output values), and a range (the set of all actual output values). Both the codomain and range depend on the domain. For an abstract function, the domain is usually viewed as the largest reasonable collection of input values; for a function that models a physical phenomenon, the domain is usually determined by the context of possibilities for the input in the phenomenon being considered.

1.2.5 Exercises

1. Based on the graphs of $f(x)$ and $g(x)$ below, answer the following questions.

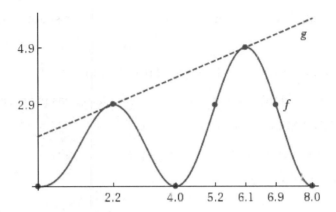

(a) Find $f(5.2)$.

(b) Fill in the blanks in each of the two points below to correctly complete the coordinates of two points on the graph of $g(x)$.

(6.1 , _____)   (_____ , 2.9)

(c) For what value(s) of x is/are $f(x) = 2.9$?

(d) For what value(s) of x is/are $f(x) = g(x)$?

2. The table below $A = f(d)$, the amount of money A (in billions of dollars) in bills of denomination d circulating in US currency in 2005. For example according to the table values below there were \$60.2 billion worth of \$50 bills in circulation.

Denomination (value of bill)	1	5	10	20	50	100
Dollar Value in Circulation	8.4	9.7	14.8	110.1	60.2	524.5

a) Find $f(20)$.

b) Using your answer in (a), what was the total *number of \$20 bills* (not amount of money) in circulation in 2005?

c) Are the following statements True or False?

   (i) There were more 20 dollar bills than 100 dollar bills

   (ii) There were more 5 dollar bills than 20 dollar bills

3. Let $f(t)$ denote the number of people eating in a restaurant t minutes after 5 PM. Answer the following questions:

a) Which of the following statements best describes the significance of the expression $f(4) = 21$?

   ⊙ There are 4 people eating at 5:21 PM

   ⊙ There are 21 people eating at 5:04 PM

   ⊙ There are 21 people eating at 9:00 PM

   ⊙ Every 4 minutes, 21 more people are eating

   ⊙ None of the above

b) Which of the following statements best describes the significance of the expression $f(a) = 20$?

⊙ a minutes after 5 PM there are 20 people eating

⊙ Every 20 minutes, the number of people eating has increased by a people

⊙ At 5:20 PM there are a people eating

⊙ a hours after 5 PM there are 20 people eating

⊙ None of the above

c) Which of the following statements best describes the significance of the expression $f(20) = b$?

⊙ Every 20 minutes, the number of people eating has increased by b people

⊙ b minutes after 5 PM there are 20 people eating

⊙ At 5:20 PM there are b people eating

⊙ b hours after 5 PM there are 20 people eating

⊙ None of the above

d) Which of the following statements best describes the significance of the expression $n = f(t)$?

⊙ Every t minutes, n more people have begun eating

⊙ n hours after 5 PM there are t people eating

⊙ n minutes after 5 PM there are t people eating

⊙ t hours after 5 PM there are n people eating

⊙ None of the above

4. Chicago's average monthly rainfall, $R = f(t)$ inches, is given as a function of the month, t, where January is $t = 1$, in the table below.

t, month	1	2	3	4	5	6	7	8
R, inches	1.8	1.8	2.7	3.1	3.5	3.7	3.5	3.4

(a) Solve $f(t) = 3.4$.

The solution(s) to $f(t) = 3.4$ can be interpreted as saying

⊙ Chicago's average rainfall is least in the month of August.

⊙ Chicago's average rainfall in the month of August is 3.4 inches.

⊙ Chicago's average rainfall increases by 3.4 inches in the month of May.

⊙ Chicago's average rainfall is greatest in the month of May.

⊙ None of the above

(b) Solve $f(t) = f(5)$.

The solution(s) to $f(t) = f(5)$ can be interpreted as saying

⊙ Chicago's average rainfall is greatest in the month of May.

⊙ Chicago's average rainfall is 3.5 inches in the months of May and July.

⊙ Chicago's average rainfall is 3.5 inches in the month of May.

⊙ Chicago's average rainfall is 3.5 inches in the month of July.

⊙ None of the above

5. A national park records data regarding the total fox population F over a 12 month pe- riod, where $t = 0$ means January 1, $t = 1$ means February 1, and so on. Below is the table of values they recorded:

t, month	0	1	2	3	4	5	6	7	8	9	10	11
F, foxes	150	143	125	100	75	57	50	57	75	100	125	143

(a) Is t a function of F?

(b) Let $g(t) = F$ denote the fox population in month t. Find all solution(s) to the equation $g(t) = 125$. If there is more than one solution, give your answer as a comma separated list of numbers.

6. An open box is to be made from a flat piece of material 20 inches long and 6 inches wide by cutting equal squares of length x from the corners and folding up the sides.

Write the volume V of the box as a function of x. Leave it as a product of factors, do not multiply out the factors.

If we write the domain of the box as an open interval in the form (a, b), then what is a and what is b?

7. Consider an inverted conical tank (point down) whose top has a radius of 3 feet and that is 2 feet deep. The tank is initially empty and then is filled at a constant rate of 0.75 cubic feet per minute. Let $V = f(t)$ denote the volume of water (in cubic feet) at time t in minutes, and let $h = g(t)$ denote the depth of the water (in feet) at time t.

   a. Recall that the volume of a conical tank of radius r and depth h is given by the formula $V = \frac{1}{3}\pi r^2 h$. How long will it take for the tank to be completely full and how much water will be in the tank at that time?

   b. On the provided axes, sketch possible graphs of both $V = f(t)$ and $h = g(t)$, making them as accurate as you can. Label the scale on your axes and points whose coordinates you know for sure; write at least one sentence for each graph to discuss the shape of your graph and why it makes sense in the context of the model.

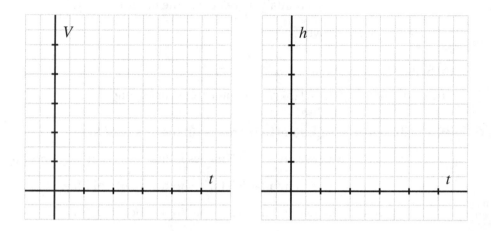

c. What is the domain of the model $h = g(t)$? its range? why?

d. It's possible to show that the formula for the function g is $g(t) = \left(\frac{t}{\pi}\right)^{1/3}$. Use a computational device to generate two plots: on the axes at left, the graph of the model $h = g(t) = \left(\frac{t}{\pi}\right)^{1/3}$ on the domain that you decided in (c); on the axes at right, the graph of the abstract function $y = p(x) = \left(\frac{t}{\pi}\right)^{1/3}$ on a wider domain than that of g. What are the domain and range of p and how do these differ from those of the physical model g?

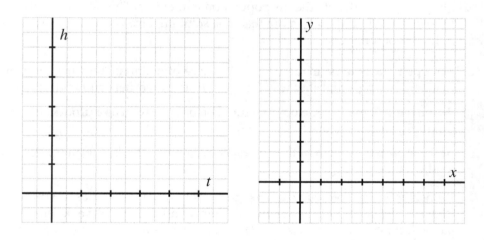

8. A person is taking a walk along a straight path. Their velocity, v (in feet per second), which is a function of time t (in seconds), is given by the graph in Figure 1.2.19.

    a. What is the person's velocity when $t = 2$? when $t = 7$?

    b. Are there any times when the person's velocity is exactly $v = 3$ feet per second? If yes, identify all such times; if not, explain why.

    c. Describe the person's behavior on the time interval $4 \le t \le 5$.

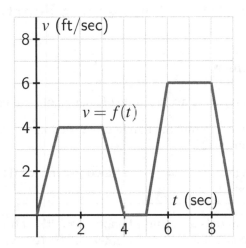

Figure 1.2.19: The velocity graph for a person walking along a straight path.

d. On which time interval does the person travel a farther distance: $[1, 3]$ or $[6, 8]$? Why?

9. A driver of a new car periodically keeps track of the number of gallons of gas remaining in their car's tank, while simultaneously tracking the trip odometer mileage. Their data is recorded in the following table. Note that at mileages where they add fuel to the tank, they record the mileage twice: once before fuel is added, and once afterward.

D (miles)	0	50	100	100	150	200	250	300	300	350
G (gallons)	4.5	3.0	1.5	10.0	8.5	7.0	5.5	4.0	11.0	9.5

Table 1.2.20: Remaining gas as a function of distance traveled.

Use the table to respond to the questions below.

a. Can the amoung of fuel in the gas tank, G, be viewed as a function of distance traveled, D? Why or why not?

b. Does the car's fuel economy appear to be constant or does it appear to vary? Why?

c. At what odometer reading did the driver put the most gas in the tank?

1.3 The Average Rate of Change of a Function

Motivating Questions

- What do we mean by the average rate of change of a function on an interval?

- What does the average rate of change of a function measure? How do we interpret its meaning in context?

- How is the average rate of change of a function connected to a line that passes through two points on the curve?

Given a function that models a certain phenomenon, it's natural to ask such questions as "how is the function changing on a given interval" or "on which interval is the function changing more rapidly?" The concept of *average rate of change* enables us to make these questions more mathematically precise. Initially, we will focus on the average rate of change of an object moving along a straight-line path.

For a function s that tells the location of a moving object along a straight path at time t, we define the average rate of change of s on the interval $[a, b]$ to be the quantity

$$AV_{[a,b]} = \frac{s(b) - s(a)}{b - a}.$$

Note particularly that the average rate of change of s on $[a, b]$ is measuring the *change in position* divided by the *change in time*.

> **Preview Activity 1.3.1.** Let the height function for a ball tossed vertically be given by $s(t) = 64 - 16(t - 1)^2$, where t is measured in seconds and s is measured in feet above the ground.
>
> a. Compute the value of $AV_{[1.5,2.5]}$.
>
> b. What are the units on the quantity $AV_{[1.5,2.5]}$? What is the meaning of this number in the context of the rising/falling ball?
>
> c. In *Desmos*, plot the function $s(t) = 64 - 16(t-1)^2$ along with the points $(1.5, s(1.5))$ and $(2.5, s(2.5))$. Make a copy of your plot on the axes in Figure 1.3.1, labeling key points as well as the scale on your axes. What is the domain of the model? The range? Why?
>
> d. Work by hand to find the equation of the line through the points $(1.5, s(1.5))$ and $(2.5, s(2.5))$. Write the line in the form $y = mt + b$ and plot the line in *Desmos*, as well as on the axes above.
>
> e. What is a geometric interpretation of the value $AV_{[1.5,2.5]}$ in light of your work in the preceding questions?

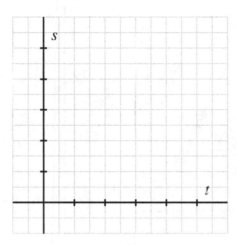

Figure 1.3.1: Axes for plotting the position function.

    f. How do your answers in the preceding questions change if we instead consider the interval $[0.25, 0.75]$? $[0.5, 1.5]$? $[1, 3]$?

1.3.1 Defining and interpreting the average rate of change of a function

In the context of a function that measures height or position of a moving object at a given time, the meaning of the average rate of change of the function on a given interval is the *average velocity* of the moving object because it is the ratio of *change in position* to *change in time*. For example, in Preview Activity 1.3.1, the units on $AV_{[1.5, 2.5]} = -32$ are "feet per second" since the units on the numerator are "feet" and on the denominator "seconds". Morever, -32 is numerically the same value as the slope of the line that connects the two corresponding points on the graph of the position function, as seen in Figure 1.3.2. The fact that the average rate of change is negative in this example indicates that the ball is falling.

While the average rate of change of a position function tells us the moving object's average velocity, in other contexts, the average rate of change of a function can be similarly defined and has a related interpretation. We make the following formal definition.

Definition 1.3.4 For a function f defined on an interval $[a, b]$, the **average rate of change of f on $[a, b]$** is the quantity

$$AV_{[a,b]} = \frac{f(b) - f(a)}{b - a}.$$

◊

In every situation, the units on the average rate of change help us interpret its meaning, and those units are always "units of output per unit of input." Moreover, the average rate of change of f on $[a, b]$ always corresponds to the slope of the line between the points $(a, f(a))$ and $(b, f(b))$, as seen in Figure 1.3.3.

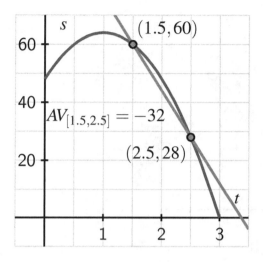 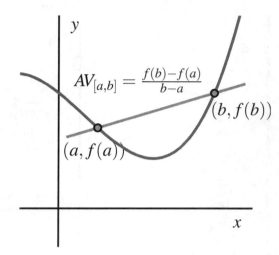

Figure 1.3.2: The average rate of change of s on $[1.5, 2.5]$ for the function in Preview Activity 1.3.1.

Figure 1.3.3: The average rate of change of an abstract function f on the interval $[a, b]$.

Activity 1.3.2. According to the US census, the populations of Kent and Ottawa Counties in Michigan where GVSU is located[1] from 1960 to 2010 measured in 10-year intervals are given in the following tables.

1960	1970	1980	1990	2000	2010
363,187	411,044	444,506	500,631	574,336	602,622

Table 1.3.5: Kent County population data.

1960	1970	1980	1990	2000	2010
98,719	128,181	157,174	187,768	238,313	263,801

Table 1.3.6: Ottawa county population data.

Let $K(Y)$ represent the population of Kent County in year Y and $W(Y)$ the population of Ottawa County in year Y.

    a. Compute $AV_{[1990,2010]}$ for both K and W.

    b. What are the units on each of the quantities you computed in (a.)?

    c. Write a careful sentence that explains the meaning of the average rate of change of the Ottawa county population on the time interval $[1990, 2010]$. Your sentence should begin something like "In an average year between 1990 and 2010, the population of Ottawa County was ..."

d. Which county had a greater average rate of change during the time interval [2000, 2010]? Were there any intervals in which one of the counties had a negative average rate of change?

e. Using the given data, what do you predict will be the population of Ottawa County in 2018? Why?

The average rate of change of a function on an interval gives us an excellent way to describe how the function behaves, on average. For instance, if we compute $AV_{[1970,2000]}$ for Kent County, we find that

$$AV_{[1970,2000]} = \frac{574,336 - 411,044}{30} = 5443.07,$$

which tells us that in an average year from 1970 to 2000, the population of Kent County increased by about 5443 people. Said differently, we could also say that from 1970 to 2000, Kent County was growing at an average rate of 5443 people per year. These ideas also afford the opportunity to make comparisons over time. Since

$$AV_{[1990,2000]} = \frac{574,336 - 500,631}{10} = 7370.5,$$

we can not only say that Kent county's population increased by about 7370 in an average year between 1990 and 2000, but also that the population was growing faster from 1990 to 2000 than it did from 1970 to 2000.

Finally, we can even use the average rate of change of a function to predict future behavior. Since the population was changing on average by 7370.5 people per year from 1990 to 2000, we can estimate that the population in 2002 is

$$K(2002) \approx K(2000) + 2 \cdot 7370.5 = 574,336 + 14,741 = 589,077.$$

1.3.2 How average rate of change indicates function trends

We have already seen that it is natural to use words such as "increasing" and "decreasing" to describe a function's behavior. For instance, for the tennis ball whose height is modeled by $s(t) = 64 - 16(t - 1)^2$, we computed that $AV_{[1.5,2.5]} = -32$, which indicates that on the interval [1.5, 2.5], the tennis ball's height is decreasing at an average rate of 32 feet per second. Similarly, for the population of Kent County, since $AV_{[1990,2000]} = 7370.5$, we know that on the interval [1990, 2000] the population is increasing at an average rate of 7370.5 people per year.

We make the following formal definitions to clarify what it means to say that a function is increasing or decreasing.

---

[1]Grand Rapids is in Kent, Allendale in Ottawa.

Definition 1.3.7 Let f be a function defined on an interval (a, b) (that is, on the set of all x for which $a < x < b$). We say that f is **increasing on** (a, b) provided that the function is always rising as we move from left to right. That is, for any x and y in (a, b), if $x < y$, then $f(x) < f(y)$.

Similarly, we say that f is **decreasing on** (a, b) provided that the function is always falling as we move from left to right. That is, for any x and y in (a, b), if $x < y$, then $f(x) > f(y)$. ◊

If we compute the average rate of change of a function on an interval, we can decide if the function is increasing or decreasing *on average* on the interval, but it takes more work[2] to decide if the function is increasing or decreasing *always* on the interval.

> **Activity 1.3.3.** Let's consider two different functions and see how different computations of their average rate of change tells us about their respective behavior. Plots of q and h are shown in Figures 1.3.8 and 1.3.9.
>
> a. Consider the function $q(x) = 4 - (x - 2)^2$. Compute $AV_{[0,1]}$, $AV_{[1,2]}$, $AV_{[2,3]}$, and $AV_{[3,4]}$. What do your last two computations tell you about the behavior of the function q on $[2, 4]$?
>
> b. Consider the function $h(t) = 3 - 2(0.5)^t$. Compute $AV_{[-1,1]}$, $AV_{[1,3]}$, and $AV_{[3,5]}$. What do your computations tell you about the behavior of the function h on $[-1, 5]$?
>
> c. On the graphs in Figures 1.3.8 and 1.3.9, plot the line segments whose respective slopes are the average rates of change you computed in (a) and (b).

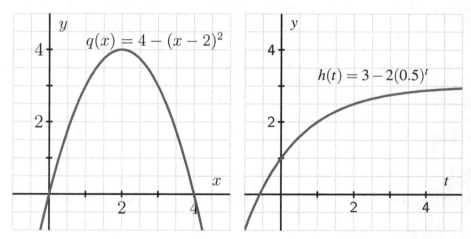

Figure 1.3.8: Plot of q from part (a).   **Figure 1.3.9:** Plot of h from part (b).

> d. True or false: Since $AV_{[0,3]} = 1$, the function q is increasing on the interval $(0, 3)$. Justify your decision.
>
> e. Give an example of a function that has the same average rate of change no matter what interval you choose. You can provide your example through a table, a graph, or a formula; regardless of your choice, write a sentence to explain.

---

[2]Calculus offers one way to justify that a function is always increasing or always decreasing on an interval.

It is helpful be able to connect information about a function's average rate of change and its graph. For instance, if we have determined that $AV_{[-3,2]} = 1.75$ for some function f, this tells us that, on average, the function rises between the points $x = -3$ and $x = 2$ and does so at an average rate of 1.75 vertical units for every horizontal unit. Moreover, we can even determine that the difference between $f(2)$ and $f(-3)$ is

$$f(2) - f(-3) = 1.75 \cdot 5 = 8.75$$

since $\frac{f(2)-f(-3)}{2-(-3)} = 1.75.$

Activity 1.3.4. Sketch at least two different possible graphs that satisfy the criteria for the function stated in each part. Make your graphs as significantly different as you can. If it is impossible for a graph to satisfy the criteria, explain why.

a. f is a function defined on $[-1, 7]$ such that $f(1) = 4$ and $AV_{[1,3]} = -2$.

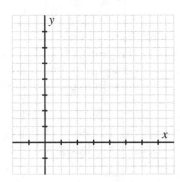
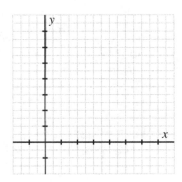

b. g is a function defined on $[-1, 7]$ such that $g(4) = 3$, $AV_{[0,4]} = 0.5$, and g is not always increasing on $(0, 4)$.

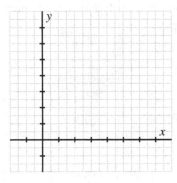
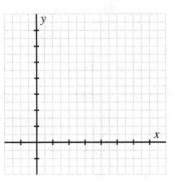

c. h is a function defined on $[-1,7]$ such that $h(2) = 5$, $h(4) = 3$ and $AV_{[2,4]} = -2$.

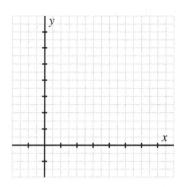 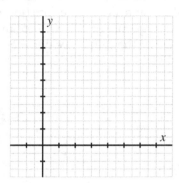

1.3.3 Summary

- For a function f defined on an interval $[a,b]$, the average rate of change of f on $[a,b]$ is the quantity

$$AV_{[a,b]} = \frac{f(b) - f(a)}{b - a}.$$

- The value of $AV_{[a,b]} = \frac{f(b)-f(a)}{b-a}$ tells us how much the function rises or falls, on average, for each additional unit we move to the right on the graph. For instance, if $AV_{[3,7]} = 0.75$, this means that for additional 1-unit increase in the value of x on the interval $[3,7]$, the function increases, on average, by 0.75 units. In applied settings, the units of $AV_{[a,b]}$ are "units of output per unit of input".

- The value of $AV_{[a,b]} = \frac{f(b)-f(a)}{b-a}$ is also the slope of the line that passes through the points $(a, f(a))$ and $(b, f(b))$ on the graph of f, as shown in Figure 1.3.3.

1.3.4 Exercises

1. Let P_1 and P_2 be the populations (in hundreds) of Town 1 and Town 2, respectively. The table below shows data for these two populations for five different years.

Year	1980	1983	1987	1993	1999
P_1	49	53	57	61	65
P_2	79	72	65	58	51

Find the average rate of change of each population over each of the time intervals below.

(a) From 1980 to 1987, the average rate of change of the population of Town 1 was _____ hundred people per year, and the average rate of change of the population of Town 2 was _____ hundred people per year.

(b) From 1987 to 1999, the average rate of change of the population of Town 1 was _____ hundred people per year, and the average rate of change of the population of Town 2

was ____ hundred people per year.

(c) From 1980 to 1999, the average rate of change of the population of Town 1 was ____ hundred people per year, and the average rate of change of the population of Town 2 was ____ hundred people per year.

2. *(a)* What is the average rate of change of $g(x) = -6 - 5x$ between the points $(-4, 14)$ and $(5, -31)$?

    (b) The function g is   ($\square$ increasing $\square$ decreasing)   on the interval $-4 \le x \le 5$.

3. Find the average rate of change of $f(x) = 3x^2 + 7$ between each of the pairs of points below.

    (a) Between $(3, 34)$ and $(5, 82)$

    (b) Between (c, k) and (q, t)

    (c) Between $(x, \ f(x))$ and $(x + h, \ f(x + h))$

4. In 2005, you have 45 CDs in your collection. In 2008, you have 130 CDs. In 2012, you have 50 CDs. What is the average rate of change in the size of your CD collection between:

    (a) 2005 and 2008? _____

    (b) 2008 and 2012? _____

    (c) 2005 and 2012? _____

5. Based on the graphs of $f(x)$ and $g(x)$ below, answer the following questions. You should not approximate any of your answers.

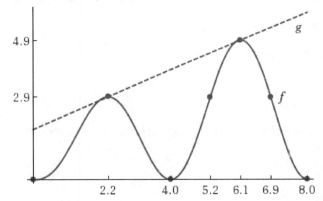

    a) What is the average rate of change of $f(x)$ over the interval $2.2 \le x \le 6.1$? _____

    b) What is the average rate of change of $g(x)$ over the interval $2.2 \le x \le 6.1$? _____

6. The graph below shows the distance traveled, D (in miles) as a function of time, t (in hours).

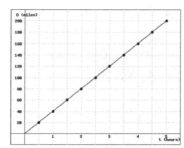

a) For each of the intervals, find the values of ΔD and Δt between the indicated start and end times. Enter your answers in their respective columns in the table below.

Time Interval	ΔD	Δt
t = 1.5 to t = 4.5		
t = 2 to t = 4.5		
t = 1 to t = 3		

b) Based on your results from (a) it follows that the average rate of change of D is constant, it does not depend over which interval of time you choose. What is the constant rate of change of D?

c) Which of the statements below CORRECTLY explains the significance of your answer to part (b)? Select ALL that apply (more than one may apply).

□ It is the average velocity of the car over the first two hours.

□ It is the total distance the car travels in five hours.

□ It is how far the car will travel in a half-hour.

□ It represents the car's velocity.

□ It is the acceleration of the car over the five hour time interval.

□ It is the slope of the line.

□ None of the above

7. Let $f(x) = 36 - x^2$.

a) Compute each of the following expressions and interpret each as an average rate of change:

(i) $\frac{f(4)-f(0)}{4-0} = $ _____

(ii) $\frac{f(6)-f(4)}{6-4} = $ _____

(iii) $\frac{f(6)-f(0)}{6-0} = $ _____

b) Based on the graph sketched below, match each of your answers in (i) - (iii) with one of the lines labeled A - F. Type the corresponding letter of the line segment next to the appropriate formula. Clearly not all letters will be used.

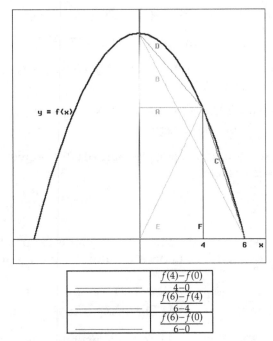

	$\dfrac{f(4)-f(0)}{4-0}$
	$\dfrac{f(6)-f(4)}{6-4}$
	$\dfrac{f(6)-f(0)}{6-0}$

8. The table below gives the average temperature, T, at a depth d, in a borehole in Belleterre, Quebec.

d, depth (m)	T, temp $(C°)$
25	5.50
50	5.20
75	5.10
100	5.10
125	5.30
150	5.50
175	5.75
200	6.00
225	6.25
250	6.50
275	6.75
300	7.00

Evaluate $\Delta T/\Delta d$ on the following intervals

a) $125 \le d \le 275$ $\Delta T/\Delta d =$ _____

b) $25 \le d \le 125$ $\Delta T/\Delta d =$ _____

c) $75 \le d \le 200$ $\Delta T/\Delta d =$ _____

d) Which of the statements below correctly explains the significance of your answer to part (c)? Select all that apply (more than one may apply).

☐ On average, the temperature is changing at a rate of 0.0072 degrees Celsius per

35

minute over the interval $75 \le d \le 200$.

☐ 0.0072 is the slope of the graph of at $d = 75$.

☐ The temperature changes by a total of 0.0072 degrees Celsius when moving from a depth 75 meters to 200 meters.

☐ Over the interval from 75 meters to 200 meters, the temperature changes on average at a rate of 0.0072 degrees Celsius per meter.

☐ The temperature is changing at a rate of 0.0072 degrees Celsius per minute when the depth is 75 meters.

☐ None of the above

9. A cold can of soda is removed from a refrigerator. Its temperature F in degrees Fahrenheit is measured at 5-minute intervals, as recorded in the following table.

t (minutes)	0	5	10	15	20	25	30	35
F (Fahrenheit temp)	37.00	44.74	50.77	55.47	59.12	61.97	64.19	65.92

Table 1.3.10: Data for the soda's temperature as a function of time.

a. Determine $AV_{[0,5]}$, $AV_{[5,10]}$, and $AV_{[10,15]}$, including appropriate units. Choose one of these quantities and write a careful sentence to explain its meaning. Your sentence might look something like "On the interval ..., the temperature of the soda is ... on average by ... for each 1-unit increase in ...".

b. On which interval is there more total change in the soda's temperature: $[10, 20]$ or $[25, 35]$?

c. What can you observe about when the soda's temperature appears to be changing most rapidly?

d. Estimate the soda's temperature when $t = 37$ minutes. Write at least one sentence to explain your thinking.

10. The position of a car driving along a straight road at time t in minutes is given by the function $y = s(t)$ that is pictured in Figure 1.3.11. The car's position function has units measured in thousands of feet. For instance, the point $(2, 4)$ on the graph indicates that after 2 minutes, the car has traveled 4000 feet.

a. In everyday language, describe the behavior of the car over the provided time interval. In particular, carefully discuss what is happening on each of the time intervals $[0, 1]$, $[1, 2]$, $[2, 3]$, $[3, 4]$, and $[4, 5]$, plus provide commentary overall on what the car is doing on the interval $[0, 12]$.

b. Compute the average rate of change of s on the intervals $[3, 4]$, $[4, 6]$, and $[5, 8]$. Label your results using the notation "$AV_{[a,b]}$" appropriately, and include units on each quantity.

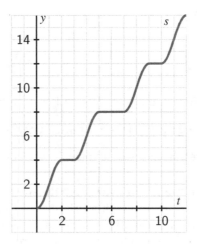

Figure 1.3.11: The graph of $y = s(t)$, the position of the car (measured in thousands of feet from its starting location) at time t in minutes.

    c. On the graph of s, sketch the three lines whose slope corresponds to the values of $AV_{[3,4]}$, $AV_{[4,6]}$, and $AV_{[5,8]}$ that you computed in (b).

    d. Is there a time interval on which the car's average velocity is 5000 feet per minute? Why or why not?

    e. Is there ever a time interval when the car is going in reverse? Why or why not?

11.   Consider an inverted conical tank (point down) whose top has a radius of 3 feet and that is 2 feet deep. The tank is initially empty and then is filled at a constant rate of 0.75 cubic feet per minute. Let $V = f(t)$ denote the volume of water (in cubic feet) at time t in minutes, and let $h = g(t)$ denote the depth of the water (in feet) at time t. It turns out that the formula for the function g is $g(t) = \left(\frac{t}{\pi}\right)^{1/3}$.

    a. In everyday language, describe how you expect the height function $h = g(t)$ to behave as time increases.

    b. For the height function $h = g(t) = \left(\frac{t}{\pi}\right)^{1/3}$, compute $AV_{[0,2]}$, $AV_{[2,4]}$, and $AV_{[4,6]}$. Include units on your results.

    c. Again working with the height function, can you determine an interval $[a, b]$ on which $AV_{[a,b]} = 2$ feet per minute? If yes, state the interval; if not, explain why there is no such interval.

    d. Now consider the volume function, $V = f(t)$. Even though we don't have a formula for f, is it possible to determine the average rate of change of the volume function on the intervals $[0, 2]$, $[2, 4]$, and $[4, 6]$? Why or why not?

1.4 Linear Functions

Motivating Questions

- What behavior of a function makes its graph a straight line?

- For a function whose graph is a straight line, what structure does its formula have?

- How can we interpret the slope of a linear function in applied contexts?

Functions whose graphs are straight lines are both the simplest and the most important functions in mathematics. Lines often model important phenomena, and even when they don't directly model phenomena, lines can often approximate other functions that do. Whether a function's graph is a straight line or not is connected directly to its average rate of change.

Preview Activity 1.4.1.

a. Let $y = f(x) = 7 - 3x$. Determine $AV_{[-3,-1]}$, $AV_{[2,5]}$, and $AV_{[4,10]}$ for the function f.

b. Let $y = g(x)$ be given by the data in Table 1.4.1.

x	−5	−4	−3	−2	−1	0	1	2	3	4	5
$g(x)$	−2.75	−2.25	−1.75	−1.25	−0.75	−0.25	0.25	0.75	1.25	1.75	2.25

Table 1.4.1: A table that defines the function $y = g(x)$.

Determine $AV_{[-5,-2]}$, $AV_{[-1,1]}$, and $AV_{[0,4]}$ for the function g.

c. Consider the function $y = h(x)$ defined by the graph in Figure 1.4.2.

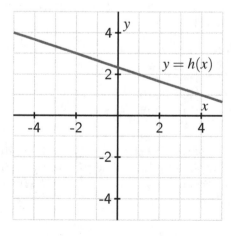

Figure 1.4.2: The graph of $y = h(x)$.

Determine $AV_{[-5,-2]}$, $AV_{[-1,1]}$, and $AV_{[0,4]}$ for the function g.

d. What do all three examples above have in common? How do they differ?

e. For the function $y = f(x) = 7 - 3x$ from (a), find the simplest expression you can for

$$AV_{[a,b]} = \frac{f(b) - f(a)}{b - a}$$

where $a \neq b$.

1.4.1 Properties of linear functions

In Preview Activity 1.4.1, we considered three different functions for which the average rate of change of each appeared to always be constant. For the first function in the preview activity, $y = f(x) = 7 - 3x$, we can compute its average rate of change on an arbitrary interval $[a, b]$. Doing so, we notice that

$$
\begin{aligned}
AV_{[a,b]} &= \frac{f(b) - f(a)}{b - a} \\
&= \frac{(7 - 3b) - (7 - 3a)}{b - a} \\
&= \frac{7 - 3b - 7 + 3a}{b - a} \\
&= \frac{-3b + 3a}{b - a} \\
&= \frac{-3(b - a)}{b - a} \\
&= -3.
\end{aligned}
$$

This result shows us that for the function $y = f(x) = 7 - 3x$, its average rate of change is always -3, regardless of the interval we choose. We will use the property of having constant rate of change as the defining property of a linear function.

Definition 1.4.3 A function f is **linear** provided that its average rate of change is constant on every choice of interval in its domain[1]. That is, for any inputs a and b for which $a \neq b$, it follows that

$$\frac{f(b) - f(a)}{b - a} = m$$

for some fixed constant m. We call m the **slope** of the linear function f. ◊

From prior study, we already know a lot about linear functions. In this section, we work to understand some familiar properties in light of the new perspective of Definition 1.4.3.

Let's suppose we know that a function f is linear with average rate of change $AV_{[a,b]} = m$ and that we also know the function value is y_0 at some fixed input x_0. That is, we know that $f(x_0) = y_0$. From this information, we can find the formula for $y = f(x)$ for *any* input

---

[1]Here we are considering functions whose domain is the set of all real numbers.

x. Working with the known point $(x_0, f(x_0))$ and any other point $(x, f(x))$ on the function's graph, we know that the average rate of change between these two points must be the constant m. This tells us that

$$\frac{f(x) - f(x_0)}{x - x_0} = m.$$

Since we are interested in finding a formula for $y = f(x)$, we solve this most recent equation for $f(x)$. Multiplying both sides by $(x - x_0)$, we see that

$$f(x) - f(x_0) = m(x - x_0).$$

Adding $f(x_0)$ to each side, it follows

$$f(x) = f(x_0) + m(x - x_0). \tag{1.4.1}$$

This shows that to determine the formula for a linear function, all we need to know is its average rate of change (or slope) and a single point the function passes through.

Example 1.4.4 Find a formula for a linear function f whose average rate of change is $m = -\frac{1}{4}$ and passes through the point $(-7, -5)$.

Solution. Using Equation (1.4.1) and the facts that $m = -\frac{1}{4}$ and $f(-7) = -5$ (that is, $x_0 = -7$ and $f(x_0) = -5$), we have

$$f(x) = -5 - \frac{1}{4}(x - (-7)) = 5 - \frac{1}{4}(x + 7).$$

□

Replacing $f(x)$ with y and $f(x_0)$ with y_0, we call Equation (1.4.1) the *point-slope form* of a line.

> **Point-slope form of a line.**
>
> A line with slope m (equivalently, average rate of change m) that passes through the point (x_0, y_0) has equation
> $$y = y_0 + m(x - x_0).$$

Activity 1.4.2. Find an equation for the line that is determined by the following conditions; write your answer in point-slope form wherever possible.

a. The line with slope $\frac{3}{7}$ that passes through $(-11, -17)$.

b. The line passing through the points $(-2, 5)$ and $(3, -1)$.

c. The line passing through $(4, 9)$ that is parallel to the line $2x - 3y = 5$.

d. Explain why the function f given by Table 1.4.5 appears to be linear and find a formula for $f(x)$.

x	$f(x)$
1	7
3	3
4	1
7	−5

Table 1.4.5: Data for a linear function f.

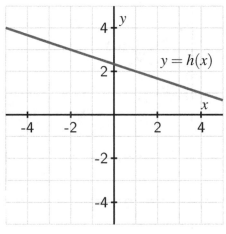

Figure 1.4.6: Plot of a linear function h.

e. Find a formula for the linear function shown in Figure 1.4.6.

Visualizing the various components of point-slope form is important. For a line through (x_0, y_0) with slope m, we know its equation is $y = y_0 + m(x - x_0)$. In Figure 1.4.7, we see that the line passes through (x_0, y_0) along with an arbitary point (x, y), which makes the vertical change between the two points given by $y - y_0$ and the horizontal change between the points $x - x_0$. This is consistent with the fact that

$$AV_{[x_0, x]} = m = \frac{y - y_0}{x - x_0}.$$

Indeed, writing $m = \frac{y - y_0}{x - x_0}$ is a rearrangement of the point-slope form of the line, $y = y_0 + m(x - x_0)$.

We naturally use the terms "increasing" and "decreasing" as from Definition 1.3.7 to describe lines based on whether their slope is positive or negative. A line with positive slope, such as the one in Figure 1.4.7, is increasing because its constant rate of change is positive, while a line with negative slope, such as in Figure 1.4.8 is decreasing because of its negative rate of change. We say that a horizontal line (one whose slope is $m = 0$) is neither increasing nor decreasing.

A special case arises when the known point on a line satisfies $x_0 = 0$. In this situation, the known point lies on the y-axis, and thus we call the point the "y-intercept" of the line. The resulting form of the line's equation is called *slope-intercept form*, which is also demonstrated in Figure 1.4.8.

> **Slope-intercept form.**
>
> For the line with slope m and passing through $(0, y_0)$, its equation is
>
> $$y = y_0 + mx.$$

Slope-intercept form follows from point-slope form from the fact that replacing x_0 with 0

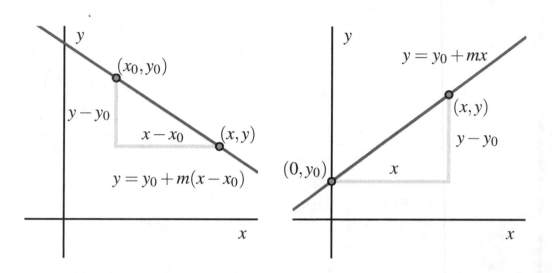

Figure 1.4.7: The point-slope form of a line's equation.

Figure 1.4.8: The slope-intercept form of a line's equation.

gives us $y = y_0 + m(x - 0) = y_0 + mx$. In many textbooks, the slope-intercept form of a line (often written $y = mx + b$) is treated as if it is the most useful form of a line. Point-slope form is actually more important and valuable since we can easily write down the equation of a line as soon as we know its slope and *any* point that lies on it, as opposed to needing to find the y-intercept, which is needed for slope-intercept form. Moreover, point-slope form plays a prominent role in calculus.

If a line is in slope-intercept or point-slope form, it is useful to be able to quickly interpret key information about the line from the form of its equation.

Example 1.4.9 For the line given by $y = -3 - 2.5(x - 5)$, determine its slope and a point that lies on the line.

Solution. This line is in point-slope form. Its slope is $m = -2.5$ and a point on the line is $(5, -3)$. □

Example 1.4.10 For the line given by $y = 6 + 0.25x$, determine its slope and a point that lies on the line.

Solution. This line is in slope-intercept form. Its slope is $m = 0.25$ and a point on the line is $(0, 6)$, which is also the line's y-intercept. □

1.4.2 Interpreting linear functions in context

Since linear functions are defined by the property that their average rate of change is constant, linear functions perfectly model quantities that change at a constant rate. In context, we can often think of slope as a rate of change; analyzing units carefully often yields significant insight.

Example 1.4.11 The Dolbear function $T = D(N) = 40 + 0.25N$ from Section 1.2 is a linear function whose slope is $m = 0.25$. What is the meaning of the slope in this context?

Solution. Recall that T is measured in degrees Fahrenheit and N in chirps per minute. We know that $m = AV_{[a,b]} = 0.25$ is the constant average rate of change of D. Its units are "units of output per unit of input", and thus "degrees Fahrenheit per chirp per minute". This tells us that the average rate of change of the temperature function is 0.25 degrees Fahrenheit per chirp per minute, which means that for each additional chirp per minute observed, we expect the temperature to rise by 0.25 degrees Fahrenheit.

Indeed, we can observe this through function values. We note that $T(60) = 55$ and $T(61) = 55.25$: one additional observed chirp per minute corresponds to a 0.25 degree increase in temperature. We also see this in the graph of the line, as seen in Figure 1.4.12: the slope between the points $(40, 50)$ and $(120, 70)$ is

$$m = \frac{70 - 50}{120 - 40}$$
$$= \frac{20}{80}$$
$$= 0.25\frac{\text{degrees F}}{\text{chirp per minute}}.$$

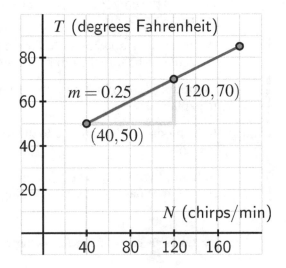

Figure 1.4.12: The linear Dolbear function with slope $m = 0.25$ degrees Fahrenheit per chirp per minute.

□

Like with the Dolbear function, it is often useful to write a linear function (whose output is called y) that models a quantity changing at a constant rate (as a function of some input t) by writing the function relationship in the form

$$y = b + mt$$

where b and m are constants. We may think of the four quantities involved in the following way:

- The constant b is the "starting value" of the output that corresponds to an input of $t = 0$;

- The constant m is the rate at which the output changes with respect to changes in the input: for each additional 1-unit change in input, the output will change by m units.

- The variable t is the independent (input) variable. A nonzero value for t corresponds to how much the input variable has changed from an initial value of 0.

- The variable y is the dependent (output) variable. The value of y results from a particular choice of t, and can be thought of as the starting output value (b) plus the change in output that results from the corresponding change in input t.

Activity 1.4.3. The summit of Africa's largest peak, Mt. Kilimanjaro[2], has two main ice fields and a glacier at its peak. Geologists measured the ice cover in the year 2000 ($t = 0$) to be approximately 1951 m^2; in the year 2007, the ice cover measured 1555 m^2.

    a. Suppose that the amount of ice cover at the peak of Mt. Kilimanjaro is changing at a constant average rate from year to year. Find a linear model $A = f(t)$ whose output is the area, A, in square meters in year t (where t is the number of years after 2000).

    b. What do the slope and A-intercept mean in the model you found in (a)? In particular, what are the units on the slope?

    c. Compute $f(17)$. What does this quantity measure? Write a complete sentence to explain.

    d. If the model holds further into the future, when do we predict the ice cover will vanish?

    e. In light of your work above, what is a reasonable domain to use for the model $A = f(t)$? What is the corresponding range?

Activity 1.4.4. In each of the following prompts, we investigate linear functions in context.

    a. A town's population initially has 28750 people present and then grows at a constant rate of 825 people per year. Find a linear model $P = f(t)$ for the number of people in the town in year t.

    b. A different town's population Q is given by the function $Q = g(t) = 42505 - 465t$. What is the slope of this function and what is its meaning in the model? Write a complete sentence to explain.

    c. A spherical tank is being drained with a pump. Initially the tank is full with $\frac{32\pi}{3}$ cubic feet of water. Assume the tank is drained at a constant rate of 1.2 cubic feet per minute. Find a linear model $V = p(t)$ for the total amount of water in the tank at time t. In addition, what is a reasonable domain for the model?

    d. A conical tank is being filled in such a way that the height of the water in the tank, h (in feet), at time t (in minutes) is given by the function $h = q(t) = 0.65t$. What can you say about how the water level is rising? Write at least one careful sentence to explain.

    e. Suppose we know that a 5-year old car's value is $10200, and that after 10 years its value is $4600. Find a linear function $C = L(t)$ whose output is the value of the car in year t. What is a reasonable domain for the model? What is the

---

[2]The main context of this problem comes from Exercise 30 on p. 27 of Connally's *Functions Modeling Change*, 5th ed.

value and meaning of the slope of the line? Write at least one careful sentence to explain.

1.4.3 Summary

- Any function f with domain all real numbers that has a constant average rate of change on every interval $[a, b]$ will have a straight line graph. We call such functions *linear* functions.

- A linear function $y = f(x)$ can be written in the form $y = f(x) = y_0 + m(x - x_0)$, where m is the slope of the line and (x_0, y_0) is a point that lies on the line. In particular, $f(x_0) = y_0$.

- In an applied context where we have a linear function that models a phenomenon in the world around us, the slope tells us the function's (constant) average rate of change. The units on the slope, m, are always "units of output per unit of input" and this enables us to articulate how the output changes in response to a 1-unit change in input.

1.4.4 Exercises

1. A town has a population of 2000 people at time $t = 0$. In each of the following cases, write a formula for the population P, of the town as a function cf year t.

   (a) The population increases by 90 people per year.

   (b) The population increases by 1 percent a year.

2. Let t be time in seconds and let $r(t)$ be the rate, in gallons per second, that water enters a reservoir:
$$r(t) = 600 - 30t.$$

   (a) Evaluate the expression $r(5)$.

   (b) Which one of the statements below best describes the physical meaning of the value of $r(5)$?

   ⊙ How many seconds until the water is entering to reservoir at a rate of 5 gallons per second.

   ⊙ The rate at which the rate of the water entering the reservoir is decreasing when 5 gallons remain in the reservoir.

   ⊙ The total amount, in gallons, of water in the reservoir afrer 5 seconds.

   ⊙ The rate, in gallons per second, at which the water is entering reservoir after 5 seconds.

   ⊙ None of the above

(c) For each of the mathematical expressions below, match one of the statements A - E below which best explains its meaning in practical terms.

  (a) The slope of the graph of $r(t)$.

  (b) The vertical intercept of the graph of $r(t)$.

  A. The rate, in gallons per second, at which the water is initially entering the reservoir.

  B. After how many seconds the water stops flowing into the reservoir and starts to drain out.

  C. The average rate, in gallons per second, at which water is flowing out of the reservoir.

  D. The initial amount of water, in gallons, in the reservoir.

  E. The rate at which the rate of water entering the reservoir is decreasing in gallons per second squared.

(d) For $0 \le t \le 30$, when does the reservoir have the most water?

(e) For $0 \le t \le 30$, when does the reservoir have the least water?

(f) If the domain of $r(t)$ is $0 \le t \le 30$, what is the range of $r(t)$?

3. A report by the US Geological Survey indicates that glaciers in Glacier National Park, Montana, are shrinking. Recent estimates indicate the area covered by glaciers has decreased from over 25.5 km^2 in 1850 to about 16.5 km^2 in 1995. Let $A = f(t)$ give the area (in square km) t years after 2000, and assume $f(t) = 16.2 - 0.062t$.

a) Find and explain the meaning of the slope. Which statement best explains its significance?

  ⊙ The area covered by glaciers is decreasing by $62,000$ m^2 every year.

  ⊙ The total area covered by glaciers decreased by 16.2 km^2 from 1850 to 2000.

  ⊙ The area covered by glaciers is decreasing by 62 m^2 every year.

  ⊙ The total area covered by glaciers is increasing by 0.062 km^2 every year.

  ⊙ The area covered by glaciers is decreasing by 16.2 km^2 every year.

  ⊙ None of the above

b) Find and explain the meaning of the A-intercept. Which statement best explains its significance?

  ⊙ The total area covered by glaciers decreased by 16.2 km^2 from 1850 to 2000.

⊙ The area covered by glaciers in 2000 was 16.2 km^2.

⊙ The total area covered by glaciers will decrease by 0.062 km^2 from 2000 to 2001.

⊙ The area covered by glaciers is decreasing by 16.2 km^2 every year.

⊙ The area covered by glaciers in 2000 was 0.062 km^2 .

⊙ None of the above

c) For both expressions listed below, enter the letter A-E of the statement which best explains their practical meaning. There are extra, unused statements.

(a) If $f(t) = 9$, then t is

(b) $f(9)$ is

A. How much area (in km^2) will be covered by glaciers in 9 years.

B. The number of years after 2000 that the amount of glacier area that has disappeared is 9 km^2.

C. The amount of glacier area (in km^2) that disappears in 9 years.

D. The number of years after 2000 that the total area covered by glaciers will be 9 km^2.

E. How much area (in km^2) will be covered by glaciers in 2009.

d) Evaluate $f(9)$.

e) How much glacier area disappears in 9 years?

f) Solve $f(t) = 9$.

4. Find a formula $p = f(h)$ for the linear equation graphed below. You can enlarge the graph by clicking on it.

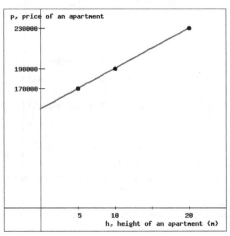

5. Find the equation for the line L (graphed in red) in the figure below. Note the x-coordinate of the point Q is 2, y-coordinate of the point P is 13, and the parabola (graphed in blue) has equation $y = x^2 + 2$.

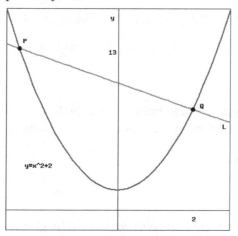

6. An apartment manager keeps careful record of how the rent charged per unit corresponds to the number of occupied units in a large complex. The collected data is shown in Table 1.4.13.[3]

Monthly Rent	$650	$700	$750	$800	$850	$900
Occupied Apartments	203	196	189	182	175	168

Table 1.4.13: Data relating occupied apartments to monthly rent.

a. Why is it reasonable to say that the number of occupied apartments is a linear function of rent?

b. Let A be the number of occupied apartments and R the monthly rent charged (in dollars). If we let $A = f(R)$, what is the slope of the linear function f? What is the meaning of the slope in the context of this question?

c. Determine a formula for $A = f(R)$. What do you think is a reasonable domain for the function? Why?

d. If the rent were to be increased to $1000, how many occupied apartments should the apartment manager expect? How much total revenue would the manager collect in a given month when rent is set at $1000?

e. Why do you think the apartment manager is interested in the data that has been collected?

7. Alicia and Dexter are each walking on a straight path. For a particular 10-second window of time, each has their velocity (in feet per second) measured and recorded as a function of time. Their respective velocity functions are plotted in Figure 1.4.14.

---

[3]This problem is a slightly modified version of one found in Carroll College's Chapter Zero resource for Active Calculus.

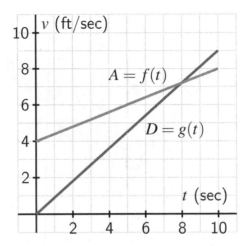

Figure 1.4.14: The velocity functions $A = f(t)$ and $D = g(t)$ for Alicia and Damon, respectively.

  a. Determine formulas for both $A = f(t)$ and $D = g(t)$.

  b. What is the value and meaning of the slope of A? Write a complete sentence to explain and be sure to include units in your response.

  c. What is the value and meaning of the average rate of change of D on the interval $[4, 8]$? Write a complete sentence to explain and be sure to include units in your response.

  d. Is there ever a time when Alicia and Damon are walking at the same velocity? If yes, determine both the time and velocity; if not, explain why.

  e. Is is possible to determine if there is ever a time when Alicia and Damon are located at the same place on the path? If yes, determine the time and location; if not, explain why not enough information is provided.

8. An inverted conical tank with depth 4 feet and radius 2 feet is completely full of water. The tank is being drained by a pump in such a way that the amount of water in the tank is decreasing at a constant rate of 1.5 cubic feet per minute. Let $V = f(t)$ denote the volume of water in the tank at time t and $h = g(t)$ the depth of the water in the tank at time t, where t is measured in minutes.

  a. How much water is in the tank at $t = 0$ when the tank is completely full?

  b. Explain why volume, V, when viewed as a function of time, t, is a linear function.

  c. Determine a formula for $V = f(t)$.

  d. At what exact time will the tank be empty?

  e. What is a reasonable domain to use for the model f? What is its corresponding range?

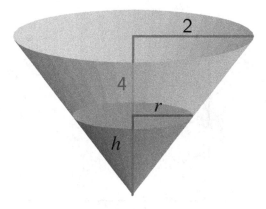

Figure 1.4.15: The inverted conical tank.

1.5 Quadratic Functions

Motivating Questions

- What patterns can we observe in how a quadratic function changes?

- What are familiar and important properties of quadratic functions?

- How can quadratic functions be used to model objects falling under the influence of gravity?

After linear functions, quadratic functions are arguably the next simplest functions in mathematics. A *quadratic function* is one that may be written in the form

$$q(x) = ax^2 - bx + c,$$

where a, b, and c are real numbers with $a \neq 0$. One of the reasons that quadratic functions are especially important is that they model the height of an object falling under the force of gravity.

> **Preview Activity 1.5.1.** A water balloon is tossed vertically from a fifth story window. Its height, h, in meters, at time t, in seconds, is modeled by the function
>
> $$h = q(t) = -5t^2 + 20t + 25.$$
>
> a. Execute appropriate computations to complete both of the following tables.
>
> | t | $h = q(t)$ |
> |---|---|
> | 0 | $q(0) = 25$ |
> | 1 | |
> | 2 | |
> | 3 | |
> | 4 | |
> | 5 | |
>
> | $[a,b]$ | $AV_{[c,b]}$ |
> |---|---|
> | $[0,1]$ | $AV_{[0,1]} = 15 \, \text{m/s}$ |
> | $[1,2]$ | |
> | $[2,3]$ | |
> | $[3,4]$ | |
> | $[4,5]$ | |
>
> **Table 1.5.1:** Function values for h at select inputs.
>
> **Table 1.5.2:** Average rates of change for h on select intervals.
>
> b. What pattern(s) do you observe in Tables 1.5.1 and 1.5.2?
>
> c. Explain why $h = q(t)$ is not a linear function. Use Definition 1.4.3 in your response.
>
> d. What is the average velocity of the water balloon in the final second before it lands? How does this value compare to the average velocity on the time interval $[4.9, 5]$?

1.5.1 Properties of Quadratic Functions

Quadratic functions are likely familiar to you from experience in previous courses. Throughout, we let $y = q(x) = ax^2 + bx + c$ where a, b, and c are real numbers with $a \neq 0$. From the outset, it is important to note that when we write $q(x) = ax^2 + bx + c$ we are thinking of an *infinite family of functions* where each member depends on the three paramaters a, b, and c.

Activity 1.5.2. Open a browser and point it to *Desmos*. In *Desmos*, enter q(x) = ax^2 + bx + c; you will be prompted to add sliders for a, b, and c. Do so. Then begin exploring with the sliders and respond to the following questions.

  a. Describe how changing the value of a impacts the graph of q.

  b. Describe how changing the value of b impacts the graph of q.

  c. Describe how changing the value of c impacts the graph of q.

  d. Which parameter seems to have the simplest effect? Which parameter seems to have the most complicated effect? Why?

  e. Is it possible to find a formula for a quadratic function that passes through the points $(0, 8)$, $(1, 12)$, $(2, 12)$? If yes, do so; if not, explain why not.

Because quadratic functions are familiar to us, we will quickly restate some of their important known properties.

Solutions to $q(x) = 0$.

Let a, b, and c be real numbers with $a \neq 0$. The equation $ax^2 + bx + c = 0$ can have 0, 1, or 2 real solutions. These real solutions are given by the quadratic formula,

$$x = \frac{-b \pm \sqrt{b^2 - 4ac}}{2a},$$

provided that $b^2 - 4ac \geq 0$.

As we can see in Figure 1.5.3, by shifting the graph of a quadratic function vertically, we can make its graph cross the x-axis 0 times (as in the graph of p), exactly 1 time (q), or twice (r). These points are the x-intercepts of the graph.

While the quadratic formula will always provide any real solutions to $q(x) = 0$, in practice it is often easier to attempt to factor before using the formula. For instance, given $q(x) = x^2 - 5x + 6$, we can find its x-intercepts quickly by factoring. Since

$$x^2 - 5x + 6 = (x - 2)(x - 3),$$

it follows that $(2, 0)$ and $(3, 0)$ are the x-intercepts of q. Note more generally that if we know the x-intercepts of a quadratic function are $(r, 0)$ and $(s, 0)$, it follows that we can write the quadratic function in the form $q(x) = a(x - r)(x - s)$.

Every quadratic function has a y-intercept; for a function of form $y = q(x) = ax^2 + bx + c$,

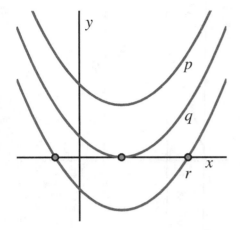

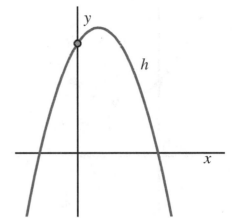

Figure 1.5.3: Three examples of quadratic functions that open up.

Figure 1.5.4: One example of a quadratic function that opens down.

the y-intercept is the point $(0, c)$, as demonstrated in Figure 1.5.4.

In addition, every quadratic function has a symmetric graph that either always curves upward or always curves downward. The graph opens upward if and only if $a > 0$ and opens downward if and only if $a < 0$. We often call the graph of a quadratic function a *parabola*. Every parabola is symmetric about a vertical line that runs through its lowest or highest point.

The vertex of a parabola.

The quadratic function $y = q(x) = ax^2 + bx + c$ has its vertex at the point $\left(-\frac{b}{2a}, q\left(-\frac{b}{2a}\right)\right)$. When $a > 0$, the vertex is the lowest point on the graph of q, while if $a < 0$, the vertex is the highest point. Moreover, the graph of q is symmetric about the vertical line $x = -\frac{b}{2a}$.

Note particularly that due to symmetry, the vertex of a quadratic function lies halfway between its x-intercepts (provided the function has x-intercepts). In both Figures 1.5.5 and 1.5.6, we see how the parabola is symmetric about the vertical line that passes through the vertex. One way to understand this symmetry can be seen by writing a given quadratic function in a different algebraic form.

Example 1.5.7 Consider the quadratic function in standard form given by $y = q(x) = 0.25x^2 - x + 3.5$. Determine constants a, h, and k so that $q(x) = a(x - h)^2 + k$, and hence determine the vertex of q. How does this alternate form of q explain the symmetry in its graph?

Solution. We first observe that we can write $q(x) = 0.25x^2 - x + 3.5$ in a form closer to $q(x) = a(x - h)^2 + k$ by factoring 0.25 from the first two terms to get

$$q(x) = 0.25(x^2 - 4x) + 3.5.$$

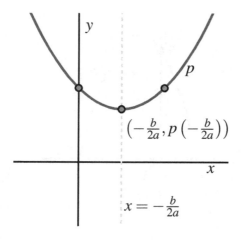

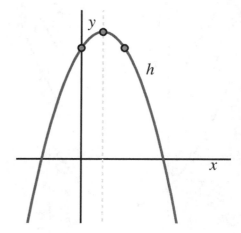

Figure 1.5.5: The vertex of a quadratic function that opens up.

Figure 1.5.6: The vertex of a quadratic function that opens down.

Next, we want to add a constant inside the parentheses to form a perfect square. Noting that $(x - 2)^2 = x^2 - 4x + 4$, we need to add 4. Since we are adding 4 inside the parentheses, the 4 is being multiplied by 0.25, which has the net effect of adding 1 to the function. To keep the function as given, we must also subtract 1, and thus we have

$$q(x) = 0.25(x^2 - 4x + 4) + 3.5 - 1.$$

It follows that $q(x) = 0.25(x - 2)^2 + 2.5$.

Next, observe that the vertex of q is $(2, 2.5)$. This holds because $(x - 2)^2$ is always greater than or equal to 0, and thus its smallest possible value is 0 when $x = 2$. Moreover, when $x = 2$, $q(2) = 2.5$.[1]

Finally, the form $q(x) = 0.25(x - 2)^2 + 2.5$ explains the symmetry of q about the line $x = 2$. Consider the two points that lie equidistant from $x = 2$ on the x-axis, z units away: $x = 2 - z$ and $x = 2 + z$. Observe that for these values,

$$q(2 - z) = 0.25(2 - z - 2)^2 + 2.5 \qquad\qquad q(2 + z) = 0.25(2 + z - 2)^2 + 2.5$$
$$= 0.25(-z)^2 + 2.5 \qquad\qquad\qquad\quad = 0.25(z)^2 + 2.5$$
$$= 0.25z^2 + 2.5 \qquad\qquad\qquad\qquad = 0.25z^2 + 2.5$$

Since $q(2 - z) = q(2 + z)$ for any choice of z, this shows the parabola is symmetric about the vertical line through its vertex. □

In Example 1.5.7, we saw some of the advantages of writing a quadratic function in the form $q(x) = a(x - h)^2 + k$. We call this the *vertex form* of a quadratic function.

---

[1]We can also verify this point is the vertex using standard form. From $q(x) = 0.25x^2 - x + 3.5$, we see that $a = 0.25$ and $b = -1$, so $x = -\frac{b}{2a} = \frac{1}{0.5} = 2$. In addition, $q(2) = 2.5$.

Vertex form of a quadratic function.

A quadratic function with vertex (h, k) may be written in the form $y = a(x - h)^2 + k$. The constant a may be determined from one other function value for an input $x \neq h$.

Activity 1.5.3. Reason algebraically using appropriate properties of quadratic functions to answer the following questions. Use *Desmos* to check your results graphically.

a. How many quadratic functions have x-intercepts at $(-5, 0)$ and $(10, 0)$ and a y-intercept at $(0, -1)$? Can you determine an exact formula for such a function? If yes, do so. If not, explain why.

b. Suppose that a quadratic function has vertex $(-3, -4)$ and opens upward. How many x-intercepts can you guarantee the function has? Why?

c. In addition to the information in (b), suppose you know that $q(-1) = -3$. Can you determine an exact formula for q? If yes, do so. If not, explain why.

d. Does the quadratic function $p(x) = -3(x + 1)^2 + 9$ have 0, 1, or 2 x-intercepts? Reason algebraically to determine the exact values of any such intercepts or explain why none exist.

e. Does the quadratic function $w(x) = -2x^2 + 10x - 20$ have 0, 1, or 2 x-intercepts? Reason algebraically to determine the exact values of any such intercepts or explain why none exist.

1.5.2 Modeling falling objects

One of the reasons that quadratic functions are so important is because of a physical fact of the universe we inhabit: for an object only being influenced by gravity, *acceleration due to gravity is constant*. If we measure time in seconds and a rising or falling object's height in feet, the gravitational constant is $g = -32$ feet per second per second.

One of the fantastic consequences of calculus — which, like the realization that acceleration due to gravity is constant, is largely due to Sir Isaac Newton in the late 1600s — is that the height of a falling object at time t is modeled by a quadratic function.

Height of an object falling under the force of gravity.

For an object tossed vertically from an initial height of s_0 feet with a velocity of v_0 feet per second, the object's height at time t (in seconds) is given by the formula

$$h(t) = -16t^2 + v_0 t + s_0$$

If height is measured instead in meters and velocity in meters per second, the gravitational constant is $g = 9.8$ and the function h has form $h(t) = -4.9t^2 + v_0 t + s_0$. (When height is measured in feet, the gravitational constant is $g = 32$.)

Activity 1.5.4. A water balloon is tossed vertically from a window at an initial height of 37 feet and with an initial velocity of 41 feet per second.

    a. Determine a formula, $s(t)$, for the function that models the height of the water balloon at time t.

    b. Plot the function in *Desmos* in an appropriate window.

    c. Use the graph to estimate the time the water balloon lands.

    d. Use algebra to find the exact time the water balloon lands.

    e. Determine the exact time the water balloon reaches its highest point and its height at that time.

    f. Compute the average rate of change of s on the intervals $[1.5, 2]$, $[2, 2.5]$, $[2.5, 3]$. Include units on your answers and write one sentence to explain the meaning of the values you found. Sketch appropriate lines on the graph of s whose respective slopes are the values of these average rates of change.

1.5.3 How quadratic functions change

So far, we've seen that quadratic functions have many interesting properties. In Preview Activity 1.5.1, we discovered an additional pattern that is particularly noteworthy.

Recall that we considered a water balloon tossed vertically from a fifth story window whose height, h, in meters, at time t, in seconds, is modeled[2] by the function

$$h = q(t) = -5t^2 + 20t + 25.$$

We then completed Table 1.5.8 and Table 1.5.9 to investigate how both function values and averages rates of change varied as we changed the input to the function.

t	$h = q(t)$
0	$q(0) = 25$
1	$q(1) = 40$
2	$q(2) = 45$
3	$q(3) = 40$
4	$q(4) = 25$
5	$q(5) = 0$

$[a, b]$	$AV_{[a,b]}$
$[0, 1]$	$AV_{[0,1]} = 15\,\text{m/s}$
$[1, 2]$	$AV_{[1,2]} = 5\,\text{m/s}$
$[2, 3]$	$AV_{[2,3]} = -5\,\text{m/s}$
$[3, 4]$	$AV_{[3,4]} = -15\,\text{m/s}$
$[4, 5]$	$AV_{[4,5]} = -25\,\text{m/s}$

Table 1.5.8: Function values for h at select inputs.　　**Table 1.5.9:** Average rates of change for h on select intervals $[a, b]$.

In Table 1.5.9, we see an interesting pattern in the average velocities of the ball. Indeed, if we remove the "AV" notation and focus on the starting value of each interval, viewing the resulting average rate of change, r, as a function of the starting value, we may consider the related table seen in Table 1.5.10, where it is apparent that r is a linear function of a.

---

[2]Here we are using $a = -5$ rather than $a = -4.9$ for simplicity.

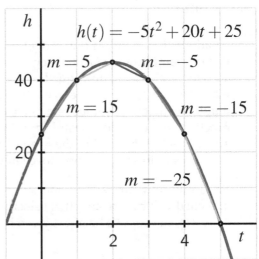

a	$r(a)$
0	$r(0) = 15$ m/s
1	$r(1) = 5$ m/s
2	$r(2) = -5$ m/s
3	$r(3) = -15$ m/s
4	$r(4) = -25$ m/s

Table 1.5.10: Data from Table 1.5.9, slightly recast.

Figure 1.5.11: Plot of $h(t) = -5t^2 + 20t + 25$ along with line segments whose slopes correspond to average rates of change.

Indeed, viewing this data graphically as in Figure 1.5.11, we observe that the average rate of change of h is itself changing in a way that seems to be represented by a linear function. While it takes key ideas from calculus to formalize this observation, for now we will simply note that for a quadratic function there seems to be a related linear function that tells us something about how the quadratic function changes. Moreover, we can also say that on the downward-opening quadratic function h that its average rate of change appears to be decreasing as we move from left to right[3].

A key closing observation here is that the fact the parabola "bends down" is apparently connected to the fact that its average rate of change decreases as we move left to right. By contrast, for a quadratic function that "bends up", we can show that its average rate of change increases as we move left to right (see Exercise 1.5.5.7). Moreover, we also see that it's possible to view the average rate of change of a function on 1-unit intervals as *itself* being a function: a process that relates an input (the starting value of the interval) to a corresponding output (the average rate of change of the original function on the resulting 1-unit interval).

For any function that consistently bends either exclusively upward or exclusively downward on a given interval (a, b), we use the following formal language[4] to describe it.

Definition 1.5.12 If a function f always bends upward on an interval (a, b), we say that f is **concave up on** (a, b). Similarly, if f always bends downward on an interval (a, b), we say that f is **concave down on** (a, b). ◊

Thus, we now call a quadratic function $q(x) = ax^2 + bx + c$ with $a > 0$ "concave up", while if $a < 0$ we say q is "concave down".

---

[3]Provided that we consider the average rate of change on intervals of the same length. Again, it takes ideas from calculus to make this observation completely precise.

[4]Calculus is needed to make Definition 1.5.12 rigorous and precise.

1.5.4 Summary

- Quadratic functions (of the form $q(x) = ax^2 + bx + c$ with $a \neq 0$) are emphatically not linear: their average rate of change is not constant, but rather depends on the interval chosen. At the same time, quadratic functions appear to change in a very regimented way: if we compute the average rate of change on several consecutive 1-unit intervals, it appears that the average rate of change itself changes at a constant rate. Quadratic functions either bend upward ($a > 0$) or bend downward ($a < 0$) and these shapes are connected to whether the average rate of change on consecutive 1-unit intervals decreases or increases as we move left to right.

- For an object with height h measured in feet at time t in seconds, if the object was launched vertically at an initial velocity of v_0 feet per second and from an initial height of s_0 feet, the object's height is given by

$$h = q(t) = -16t^2 + v_0 t + s_0.$$

That is, the object's height is completely determined by the initial height and initial velocity from which it was launched. The model is valid for the entire time until the object lands. If h is instead measured in meters and v_0 in meters per second, -16 is replaced with -4.9.

- A quadratic function q can be written in one of three familiar forms: standard, vertex, or factored[5]. Table 1.5.13 shows how, depending on the algebraic form of the function, various properties may be (easily) read from the formula. In every case, the sign of a determines whether the function opens up or opens down.

	standard	vertex	factored[6]
form	$q(x) = ax^2 + bx + c$	$q(x) = a(x - h)^2 + k$	$q(x) = a(x - r)(x - s)$
y-int	$(0, c)$	$(0, ah^2 + k)$	$(0, ars)$
x-int[7]	$\left(\frac{-b \pm \sqrt{b^2 - 4ac}}{2a}, 0\right)$	$\left(h \pm \sqrt{-\frac{k}{a}}, 0\right)$	$(r, 0), (s, 0)$
vertex	$\left(-\frac{b}{2a}, q\left(-\frac{b}{2a}\right)\right)$	(h, k)	$\left(\frac{r+s}{2}, q\left(\frac{r+s}{2}\right)\right)$

Table 1.5.13: A summary of the information that can be read from the various algebraic forms of a quadratic function

1.5.5 Exercises

1. Consider the Quadratic function $f(x) = x^2 - 5x - 24$. Find its vertex, x-intercepts, and y-intercept.

---

[5]It's not always possible to write a quadratic function in factored form involving only real numbers; this can only be done if it has 1 or 2 x-intercepts.

[6]Provided q has 1 or 2 x-intercepts. In the case of just one, we take $r = s$.

[7]Provided $b^2 - 4ac \geq 0$ for standard form; provided $-\frac{k}{a} \geq 0$ for vertex form.

2. Identify the graphs A (blue), B (red) and C (green):

_____ is the graph of the function $f(x) = (x - 3)^2$

_____ is the graph of the function $g(x) = (x + 4)^2$

_____ is the graph of the function $h(x) = x^2 - 2$

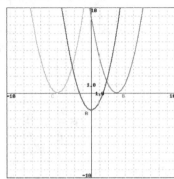

3. Find the zeros, if any, of the function $y = 4(x + 8)^2 - 8$.

4. Find the zero(s) (if any) of the function $y = x^2 - 15x + 50$

5. If a ball is thrown straight up into the air with an initial velocity of 100 ft/s, its height in feet after t second is given by $y = 100t - 16t^2$. Find the average velocity (include units,) for the time period begining when $t = 2$ seconds and lasting

(i) 0.5 seconds

(ii) 0.1 seconds

(iii) 0.01 seconds

Finally based on the above results, guess what the instantaneous velocity of the ball is when $t = 2$.

6. Two quadratic functions, f and g, are determined by their respective graphs in Figure 1.5.14.

   a. How does the information provided enable you to find a formula for f? Explain, and determine the formula.

   b. How does the information provided enable you to find a formula for g? Explain, and determine the formula.

   c. Consider an additional quadratic function h given by $h(x) = 2x^2 - 8x + 6$. Does the graph of h intersect the graph of f? If yes, determine the exact points of intersection, with justification. If not, explain why.

   d. Does the graph of h intersect the graph of g? If yes, determine the exact points of intersection, with justification. If not, explain why.

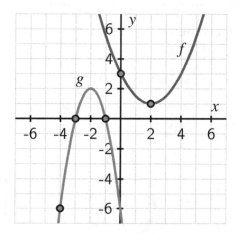

Figure 1.5.14: Two quadratic functions, f and g.

7. Consider the quadratic function f given by $f(x) = \frac{1}{2}(x-2)^2 + 1$.

    a. Determine the exact location of the vertex of f.

    b. Does f have 0, 1, or 2 x-intercepts? Explain, and determine the location(s) of any x-intercept(s) that exist.

    c. Complete the following tables of function values and average rates of change of f at the stated inputs and intervals.

x	$f(x)$
0	
1	
2	
3	
4	
5	

$[a,b]$	$AV_{[a,b]}$
$[0,1]$	
$[1,2]$	
$[2,3]$	
$[3,4]$	
$[4,5]$	

Table 1.5.16: Average rates of change for f on select intervals.

Table 1.5.15: Function values for f at select inputs.

    d. What pattern(s) do you observe in Table 1.5.15 and 1.5.16?

8. A water balloon is tossed vertically from a window on the fourth floor of a dormitory from an initial height of 56.3 feet. A person two floors above observes the balloon reach its highest point 1.2 seconds after being launched.

    a. What is the balloon's exact height at $t = 2.4$? Why?

    b. What is the exact maximum height the balloon reaches at $t = 1.2$?

    c. What exact time did the balloon land?

    d. At what initial velocity was the balloon launched?

1.6 Composite Functions

Motivating Questions

- How does the process of function composition produce a new function from two other functions?

- In the composite function $h(x) = f(g(x))$, what do we mean by the "inner" and "outer" function? What role do the domain and codomain of f and g play in determining the domain and codomain of h?

- How does the expression for $AV_{[a,a-h]}$ involve a composite function?

Recall that a function, by definition, is a process that takes a collection of inputs and produces a corresponding collection of outputs in such a way that the process produces one and only one output value for any single input value. Because every function is a process, it makes sense to think that it may be possible to take two function processes and do one of the processes first, and then apply the second process to the result.

Example 1.6.1 Suppose we know that y is a function of x according to the process defined by $y = f(x) = x^2 - 1$ and, in turn, x is a function of t via $x = g(t) = 3t - 4$. Is it possible to combine these processes to generate a new function so that y is a function of t?

Solution.   Since y depends on x and x depends on t, it follows that we can also think of y depending directly on t. We can use substitution and the notation of functions to determine this relationship.

First, it's important to realize what the rule for f tells us. In words, f says "to generate the output that corresponds to an input, take the input and square it, and then subtract 1." In symbols, we might express f more generally by writing "$f(\square) = \square^2 - 1$."

Now, observing that $y = f(x) = x^2 - 1$ and that $x = g(t) = 3t - 4$, we can substitute the expression $g(t)$ for x in f. Doing so,

$$y = f(x)$$
$$= f(g(t))$$
$$= f(3t - 4).$$

Applying the process defined by the function f to the input $3t - 4$, we see that

$$y = (3t - 4)^2 - 1,$$

which defines y as a function of t. □

When we have a situation such as in Example 1.6.1 where we use the output of one function as the input of another, we often say that we have "composed two functions". In addition, we use the notation $h(t) = f(g(t))$ to denote that a new function, h, results from composing the two functions f and g.

Preview Activity 1.6.1. Let $y = p(x) = 3x - 4$ and $x = q(t) = t^2 - 1$.

a. Let $r(t) = p(q(t))$. Determine a formula for r that depends only on t and not on p or q.

b. Recall Example 1.6.1, which involved functions similar to p and q. What is the biggest difference between your work in (a) above and in Example 1.6.1?

c. Let $t = s(z) = \frac{1}{z+4}$ and recall that $x = q(t) = t^2 - 1$. Determine a formula for $x = q(s(z))$ that depends only on z.

d. Suppose that $h(t) = \sqrt{2t^2 + 5}$. Determine formulas for two related functions, $y = f(x)$ and $x = g(t)$, so that $h(t) = f(g(t))$.

1.6.1 Composing two functions

Whenever we have two functions, say $g : A \to B$ and $f : B \to C$, where the codomain of g matches the domain of f, it is possible to link the two processes together to create a new process that we call the *composition* of f and g.

Definition 1.6.2 If f and g are functions such that $g : A \to B$ and $f : B \to C$, we define the **composition of f and g** to be the new function $h : A \to C$ given by

$$h(t) = f(g(t)).$$

We also sometimes use the notation $h = f \circ g$, where $f \circ g$ is the single function defined by $(f \circ g)(t) = f(g(t))$. ◊

We sometimes call g the "inner function" and f the "outer function". It is important to note that the inner function is actually the first function that gets applied to a given input, and then outer function is applied to the output of the inner function. In addition, in order for a composite function to make sense, we need to ensure that the range of the inner function lies within the domain of the outer function so that the resulting composite function is defined at every possible input.

In addition to the possibility that functions are given by formulas, functions can be given by tables or graphs. We can think about composite functions in these settings as well, and the following activities prompt us to consider functions given in this way.

Activity 1.6.2. Let functions p and q be given by the graphs in Figure 1.6.4 (which are each piecewise linear - that is, parts that look like straight lines are straight lines) and let f and g be given by Table 1.6.3.

x	0	1	2	3	4
$f(x)$	6	4	3	4	6
$g(x)$	1	3	0	4	2

Table 1.6.3: Table that defines f and g.

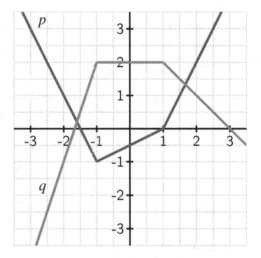

Figure 1.6.4: The graphs of p and q.

Compute each of the following quantities or explain why they are not defined.

a. $p(q(0))$

b. $q(p(0))$

c. $(p \circ p)(-1)$

d. $(f \circ g)(2)$

e. $(g \circ f)(3)$

f. $g(f(0))$

g. For what value(s) of x is $f(g(x)) = 4$?

h. For what value(s) of x is $q(p(x)) = 1$?

1.6.2 Composing functions in context

Recall Dolbear's function, $T = D(N) = 40 + 0.25N$, that relates the number of chirps per minute from a snowy cricket to the Fahrenheit temperature, T. We earlier established that D has a domain of $[40, 160]$ and a corresponding range of $[50, 85]$. In what follows, we replace T with F to emphasize that temperature is measured in Fahrenheit degrees.

The Celcius and Fahrenheit temperature scales are connected by a linear function. Indeed, the function that converts Fahrenheit to Celcius is

$$C = G(F) = \frac{5}{9}(F - 32).$$

For instance, a Fahrenheit temperature of 32 degrees corresponds to $C = G(32) = 0$ degrees Celcius.

Activity 1.6.3. Let $F = D(N) = 40 + 0.25N$ be Dolbear's function that converts an input of number of chirps per minute to degrees Fahrenheit, and let $C = G(F) = \frac{5}{9}(F - 32)$ be the function that converts an input of degrees Fahrenheit to an output of degrees Celsius.

    a. Determine a formula for the new function $H = (G \circ D)$ that depends only on the variable N.

    b. What is the meaning of the function you found in (a)?

    c. How does a plot of the function $H = (G \circ D)$ compare to that of Dolbear's function? Sketch a plot of $y = H(N) = (G \circ D)(N)$ on the blank axes to the right of the plot of Dolbear's function, and discuss the similarities and differences between them. Be sure to label the vertical scale on your axes.

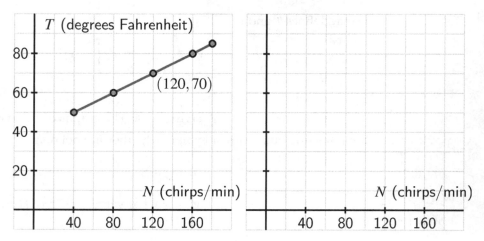

Figure 1.6.5: Dolbear's function.   Figure 1.6.6: Blank axes to plot $H = (G \circ D)(N)$.

    d. What is the domain of the function $H = G \circ D$? What is its range?

1.6.3 Function composition and average rate of change

Recall that the average rate of change of a function f on the interval $[a, b]$ is given by

$$AV_{[a,b]} = \frac{f(b) - f(a)}{b - a}.$$

In Figure 1.6.7, we see the familiar representation of $AV_{[a,b]}$ as the slope of the line joining the points $(a, f(a))$ and $(b, f(b))$ on the graph of f. In the study of calculus, we progress from the *average rate of change on an interval* to the *instantaneous rate of change of a function at a single value*; the core idea that allows us to move from an *average* rate to an *instantaneous* one is letting the interval $[a, b]$ shrink in size.

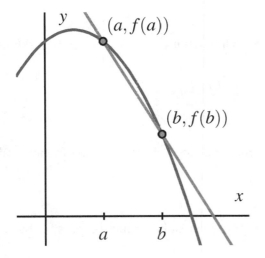

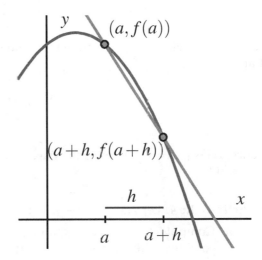

Figure 1.6.7: $AV_{[a,b]}$ is the slope of the line joining the points $(a, f(a))$ and $(b, f(b))$ on the graph of f.

Figure 1.6.8: $AV_{[a,a+h]}$ is the slope of the line joining the points $(a, f(a))$ and $(a, f(a + h))$ on the graph of f.

To think about the interval $[a, b]$ shrinking while a stays fixed, we often change our perspective and think of b as $b = a + h$, where h measures the horizontal differene from b to a. This allows us to eventually think about h getting closer and closer to 0, and in that context we consider the equivalent expression

$$AV_{[a,a+h]} = \frac{f(a + h) - f(a)}{a + h - a} = \frac{f(a + h) - f(a)}{h}$$

for the average rate of change of f on $[a, a + h]$.

In this most recent expression for $AV_{[a,a+h]}$, we see the important role that the composite function "$f(a + h)$" plays. In particular, to understand the expression for $AV_{[a,a+h]}$ we need to evaluate f at the quantity $(a + h)$.

Example 1.6.9 Suppose that $f(x) = x^2$. Determine the simplest possible expression you can find for $AV_{[3,3+h]}$, the average rate of change of f on the interval $[3, 3 + h]$.

Solution. By definition, we know that

$$AV_{[3,3+h]} = \frac{f(3 + h) - f(3)}{h}.$$

Using the formula for f, we see that

$$AV_{[3,3+h]} = \frac{(3 + h)^2 - (3)^2}{h}.$$

Expanding the numerator and combining like terms, it follows that

$$AV_{[3,3+h]} = \frac{(9 + 6h + h^2) - 9}{h}$$

$$= \frac{6h + h^2}{h}.$$

Removing a factor of h in the numerator and observing that $h \neq 0$, we can simplify and find that

$$AV_{[3,3+h]} = \frac{h(6 + h)}{h}$$

$$= 6 + h.$$

Hence, $AV_{[3,3+h]} = 6 + h$, which is the average rate of change of $f(x) = x^2$ on the interval $[3, 3 + h]$.[1]

□

Activity 1.6.4. Let $f(x) = 2x^2 - 3x + 1$ and $g(x) = \frac{5}{x}$.

a. Compute $f(1 + h)$ and expand and simplify the result as much as possible by combining like terms.

b. Determine the most simplified expression you can for the average rate of change of f on the interval $[1, 1 + h]$. That is, determine $AV_{[1,1+h]}$ for f and simplify the result as much as possible.

c. Compute $g(1 + h)$. Is there any valid algebra you can do to write $g(1 + h)$ more simply?

d. Determine the most simplified expression you can for the average rate of change of g on the interval $[1, 1 + h]$. That is, determine $AV_{[1,1+h]}$ for g and simplify the result.

In Activity 1.6.4, we see an important setting where algebraic simplification plays a crucial role in calculus. Because the expresssion

$$AV_{[a,a+h]} = \frac{f(a + h) - f(a)}{h}$$

always begins with an h in the denominator, in order to precisely understand how this quantity behaves when h gets close to 0, a simplified version of this expression is needed. For instance, as we found in part (b) of Activity 1.6.4, it's possible to show that for $f(x) = 2x^2 - 3x + 1$,

$$AV_{[1,1+h]} = 2h + 1,$$

which is a much simpler expression to investigate.

1.6.4 Summary

- When defined, the composition of two functions f and g produces a single new function $f \circ g$ according to the rule $(f \circ g)(x) = f(g(x))$. We note that g is applied first to the input x, and then f is applied to the output $g(x)$ that results from g.

---

[1]Note that $6 + h$ is a *linear* function of h. This computation is connected to the observation we made in Table 1.5.9 regarding how there's a linear aspect to how the average rate of change of a quadratic function changes as we modify the interval.

- In the composite function $h(x) = f(g(x))$, the "inner" function is g and the "outer" function is f. Note that the inner function gets applied to x first, even though the outer function appears first when we read from left to right. The composite function is only defined provided that the codomain of g matches the domain of f: that is, we need any possible outputs of g to be among the allowed inputs for f. In particular, we can say that if $g : A \rightarrow B$ and $f : B \rightarrow C$, then $f \circ g : A \rightarrow C$. Thus, the domain of the composite function is the domain of the inner function, and the codomain of the composite function is the codomain of the outer function.

- Because the expression $AV_{[a,a+h]}$ is defined by

$$AV_{[a,a+h]} = \frac{f(a+h) - f(a)}{h}$$

and this includes the quantity $f(a+h)$, the average rate of change of a function on the interval $[a, a+h]$ always involves the evaluation of a composite function expression. This idea plays a crucial role in the study of calculus.

1.6.5 Exercises

1. Suppose $r = f(t)$ is the radius, in centimeters, of a circle at time t minutes, and $A(r)$ is the area, in square centimeters, of a circle of radius r centimeters.

    Which of the following statements best explains the meaning of the composite function $A(f(t))$?

    ⊙ The area of a circle, in square centimeters, of radius r centimeters.

    ⊙ The area of a circle, in square centimeters, at time t minutes.

    ⊙ The radius of a circle, in centimeters, at time t minutes.

    ⊙ The function f of the minutes and the area.

    ⊙ None of the above

2. A swinging pendulum is constructed from a piece of string with a weight attached to the bottom. The length of the pendulum depends on how much string is let out. Suppose $L = f(t)$ is the length, in centimeters, of the pendulum at time t minutes, and $P(L)$ is the period, in seconds, of a pendulum of length L.

    Which of the following statements best explains the meaning of the composite function $P(f(t))$?

    ⊙ The period P of the pendulum, in minutes, after t minutes have elapsed.

    ⊙ The period P of the pendulum, in seconds, when the pendulum has length L meters.

    ⊙ The period P of the pendulum, in minutes, when the pendulum has length L meters.

    ⊙ The period P of the pendulum, in seconds, after t minutes have elapsed.

⊙ None of the above

3. The formula for the volume of a cube with side length s is $V = s^3$. The formula for the surface area of a cube is $A = 6s^2$.

(a) Find the formula for the function $s = f(A)$.

Which of the statements best explains the meaning of $s = f(A)$?

⊙ The side length for a cube of surface area A

⊙ The side length for a cube of volume V

⊙ The volume of a cube of side length s

⊙ The surface area of a cube of side length s

(b) If $V = g(s)$, find a formula for $g(f(A))$.

Which of the statements best explains the meaning of $g(f(A))$?

⊙ The volume for a cube of side length s

⊙ The surface area for a cube of side length s

⊙ The volume for a cube with surface area A

⊙ The surface area for a cube of volume V

4. Given that $f(x) = 5x - 6$ and $g(x) = 2x - 2$, calculate

(a) $f \circ g(x) =$ _____

(b) $g \circ f(x) =$ _____

(c) $f \circ f(x) =$ _____

(d) $g \circ g(x) =$ _____

5. This problem gives you some practice identifying how more complicated functions can be built from simpler functions.

Let $f(x) = x^3 - 27$ and let $g(x) = x - 3$. Match the functions defined below with the letters labeling their equivalent expressions.

1. $f(x)/g(x)$ A. $-27 + x^6$

2. $f(x^2)$ B. $9 + 3x + x^2$

3. $(f(x))^2$ C. $729 - 54x^3 + x^6$

4. $g(f(x))$ D. $-30 + x^3$

6. The number of bacteria in a refrigerated food product is given by $N(T) = 27T^2 - 97T + 51$, $3 < T < 33$ where T is the temperature of the food.

   When the food is removed from the refrigerator, the temperature is given by $T(t) = 4t + 1.7$, where t is the time in hours. Find the composite function $N(T(t))$.

   Find the time when the bacteria count reaches 14225.

7. Let $f(x) = 5x + 2$ and $g(x) = 4x^2 + 3x$. Find $(f \circ g)(-2)$ and $(f \circ g)(x)$.

8. Use the given information about various functions to answer the following questions involving composition.

   a. Let functions f and g be given by the graphs in Figure 1.6.10 and 1.6.11. An open circle means there is not a point at that location on the graph. For instance, $f(-1) = 1$, but $f(3)$ is not defined.

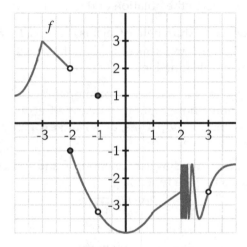
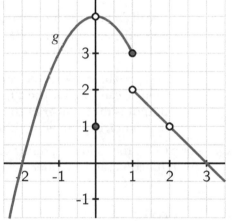

   Figure 1.6.10: Plot of $y = f(x)$.   **Figure 1.6.11:** Plot of $y = g(x)$.

   Determine $f(g(1))$ and $g(f(-2))$.

   b. Again using the functions given in (a), can you determine a value of x for which $g(f(x))$ is not defined? Why or why not?

   c. Let functions r and s be defined by Table 1.6.12.

t	-4	-3	-2	-1	0	1	2	3	4
$r(t)$	4	1	2	3	0	-3	2	-1	-4
$s(t)$	-5	-6	-7	-8	0	8	7	6	5

   Table 1.6.12: Table that defines r and s.

   Determine $(s \circ r)(3)$, $(s \circ r)(-4)$, and $(s \circ r)(a)$ for one additional value of a of your choice.

   d. For the functions r and s defined in (c), state the domain and range of each function. For how many different values of b is it possible to determine $(r \circ s)(b)$? Explain.

  e. Let $m(u) = u^3 + 4u^2 - 5u + 1$. Determine expressions for $m(x^2)$, $m(2 + h)$, and $m(a + h)$.

  f. For the function $F(x) = 4 - 3x - x^2$, determine the most simplified expression you can find for $AV_{[2,2+h]}$. Show your algebraic work and thinking fully.

9. Recall Dolbear's function that defines temperature, F, in Fahrenheit degrees, as a function of the number of chirps per minute, N, is $F = D(N) = 40 + \frac{1}{4}N$.

  a. Solve the equation $F = 40 + \frac{1}{4}N$ for N in terms of F.

  b. Say that $N = g(F)$ is the function you just found in (a). What is the meaning of this function? What does it take as inputs and what does it produce as outputs?

  c. How many chirps per minute do we expect when the outsidet temperature is 82 degrees F? How can we express this in the notation of the function g?

  d. Recall that the function that converts Fahrenheit to Celsius is $C = G(F) = \frac{5}{9}(F-32)$. Solve the equation $C = \frac{5}{9}(F - 32)$ for F in terms of C. Call the resulting function $F = p(C)$. What is the meaning of this function?

  e. Is it possible to write the chirp-rate N as a function of temperature C in Celsius? That is, can we produce a function whose input is in degrees Celsius and whose output is the number of chirps per minute? If yes, do so and explain your thinking. If not, explain why it's not possible.

10. For each of the following functions, find two simpler functions f and g such that the given function can be written as the composite function $g \circ f$.

  a. $h(x) = (x^2 + 7)^3$ c. $m(x) = \frac{1}{x^4+2x^2+1}$

  b. $r(x) = \sqrt{5 - x^3}$ d. $w(x) = 2^{3-x^2}$

11. A spherical tank has radius 4 feet. The tank is initially empty and then begins to be filled in such a way that the height of the water rises at a constant rate of 0.4 feet per minute. Let V be the volume of water in the tank at a given instant, and h the depth of the water at the same instant; let t denote the time elapsed in minutes since the tank started being filled.

  a. Calculus can be used to show that the volume, V, is a function of the depth, h, of the water in the tank according to the function

$$V = f(h) = \frac{\pi}{3}h^2(12 - h). \tag{1.6.1}$$

  What is the domain of this model? Why? What is the corresponding range?

  b. We are given the fact that the tank is being filled in such a way that the height of the water rises at a constant rate of 0.4 feet per minute. Said differently, h is a function of t whose average rate of change is constant. What kind of function does this make $h = p(t)$? Determine a formula for $p(t)$.

  c. What are the domain and range of the function $h = p(t)$? How is this tied to the dimensions of the tank?

  d. In (a) we observed that V is a function of h, and in (b) we found that h is a function of t. Use these two facts and function composition appropriately to write V as a function of t. Call the resulting function $V = q(t)$.

e. What are the domain and range of the function q? Why?

f. On the provided axes, sketch accurate graphs of $h = p(t)$ and $V = q(t)$, labeling the vertical and horizontal scale on each graph appropriately. Make your graphs as precise as you can; use a computing device to assist as needed.

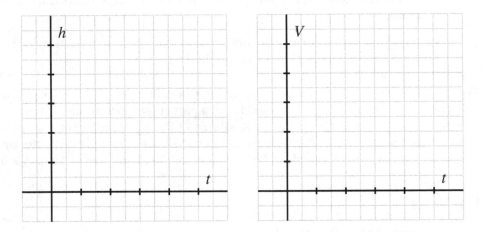

Why do each of the two graphs have their respective shapes? Write at least one sentence to explain each graph; refer explicitly to the shape of the tank and other information given in the problem.

1.7 Inverse Functions

Motivating Questions

- What does it mean to say that a given function has an inverse function?

- How can we determine whether or not a given function has a corresponding inverse function?

- When a function has an inverse function, what important properties does the inverse function have in comparison to the original function?

Because every function is a process that converts a collection of inputs to a corresponding collection of outputs, a natural question is: for a particular function, can we change perspective and think of the original function's outputs as the inputs for a reverse process? If we phrase this question algebraically, it is analogous to asking: given an equation that defines y is a function of x, is it possible to find a corresponding equation where x is a function of y?

Preview Activity 1.7.1. Recall that $F = g(C) = \frac{9}{5}C + 32$ is the function that takes Celsius temperature inputs and produces the corresponding Fahrenheit temperature outputs.

a. Show that it is possible to solve the equation $F = \frac{9}{5}C + 32$ for C in terms of F and that doing so results in the equation $C = \frac{5}{9}(F - 32)$.

b. Note that the equation $C = \frac{5}{9}(F - 32)$ expresses C as a function of F. Call this function h so that $C = h(F) = \frac{5}{9}(F - 32)$.

   Find the simplest expression that you can for the composite function $j(C) = h(g(C))$.

c. Find the simplest expression that you can for the composite function $k(F) = g(h(F))$.

d. Why are the functions j and k so simple? Explain by discussing how the functions g and h process inputs to generate outputs and what happens when we first execute one followed by the other.

1.7.1 When a function has an inverse function

In Preview Activity 1.7.1, we found that for the function $F = g(C) = \frac{9}{5}C + 32$, it's also possible to solve for C in terms of F and write $C = h(F) = \frac{5}{9}(C - 32)$. The first function, g, converts Celsius temperatures to Fahrenheit ones; the second function, h, converts Fahrenheit temperatures to Celsius ones. Thus, the process h reverses the process of g, and likewise the process of g reverses the process of h. This is also why it makes sense that $h(g(C)) = C$ and

$g(h(F)) = F$. If, for instance, we take a Celsius temperature C, convert it to Fahrenheit, and convert the result back to Celsius, we arrive back at the Celsius temperature we started with: $h(g(C)) = C$.

Similar work is sometimes possible with other functions. When we can find a new function that reverses the process of the original function, we say that the original function "has an inverse function" and make the following formal definition.

Definition 1.7.1 Let $f : A \to B$ be a function. If there exists a function $g : B \to A$ such that

$$g(f(a)) = a \text{ and } f(g(b)) = b$$

for each a in A and each b in B, then we say that f has an **inverse function** and that the function g is **the inverse of** f. ◊

Note particularly what the equation $g(f(a)) = a$ says: for any input a in the domain of f, the function g will reverse the process of f (which converts a to $f(a)$) because g converts $f(a)$ back to a.

When a given function f has a corresponding inverse function g, we usually rename g as f^{-1}, which we read aloud as "f-inverse". The equation $g(f(a)) = a$ now reads $f^{-1}(f(a)) = a$, which we interpret as saying "f-inverse converts $f(a)$ back to a". We similarly write that $f(f^{-1}(b)) = b$.

Activity 1.7.2. Recall Dolbear's function $F = D(N) = 40 + \frac{1}{4}N$ that converts the number, N, of snowy tree cricket chirps per minute to a corresponding Fahrenheit temperature. We have earlier established that the domain of D is $[40, 180]$ and the range of D is $[50, 85]$, as seen in Figure 1.2.3.

a. Solve the equation $F = 40 + \frac{1}{4}N$ for N in terms of F. Call the resulting function $N = E(F)$.

b. Explain in words the process or effect of the function $N = E(F)$. What does it take as input? What does it generate as output?

c. Use the function E that you found in (a.) to compute $j(N) = E(D(N))$. Simplify your result as much as possible. Do likewise for $k(F) = D(E(F))$. What do you notice about these two composite functions j and k?

d. Consider the equations $F = 40 + \frac{1}{4}N$ and $N = 4(F - 40)$. Do these equations express different relationships between F and N, or do they express the same relationship in two different ways? Explain.

When a given function has an inverse function, it allows us to express the same relationship from two different points of view. For instance, if $y = f(t) = 2t + 1$, we can show[1] that the function $t = g(y) = \frac{y-1}{2}$ reverses the effect of f (and vice versa), and thus $g = f^{-1}$. We observe that

$$y = f(t) = 2t + 1 \text{ and } t = f^{-1}(y) = \frac{y - 1}{2}$$

---

[1]Observe that $g(f(t)) = g(2t + 1) = \frac{(2t+1)-1}{2} = \frac{2t}{2} = t$. Similarly, $f(g(y)) = f\left(\frac{y-1}{2}\right) = 2\left(\frac{y-1}{2}\right) + 1 = y - 1 + 1 = y$.

are equivalent forms of the same equation, and thus they say the same thing from two different perspectives. The first version of the equation is solved for y in terms of t, while the second equation is solved for t in terms of y. This important principle holds in general whenever a function has an inverse function.

Two perspectives from a function and its inverse function.

If $y = f(t)$ has an inverse function, then the equations

$$y = f(t) \text{ and } t = f^{-1}(y)$$

say the exact same thing but from two different perspectives.

1.7.2 Determining whether a function has an inverse function

It's important to note in Definition 1.7.1 that we say "*If* there exists ....." That is, we don't guarantee that an inverse function exists for a given function. Thus, we might ask: how can we determine whether or not a given function has a corresponding inverse function? As with many questions about functions, there are often three different possible ways to explore such a question: through a table, through a graph, or through an algebraic formula.

Example 1.7.2 Do the functions f and g defined by Table 1.7.3 and Table 1.7.4 have corresponding inverse functions? Why or why not?

x	0	1	2	3	4
$f(x)$	6	4	3	4	6

x	0	1	2	3	4
$g(x)$	3	1	4	2	0

Table 1.7.3: The table that defines the function f.

Table 1.7.4: The table that defines the function g.

Solution. For any function, the question of whether or not it has an inverse comes down to whether or not the process of the function can be reliably reversed. For functions given in table form such as f and g, we essentially ask if it's possible to swich the input and output rows and have the new resulting table also represent a function.

The function f does not have an inverse function because there are two different inputs that lead to the same output: $f(0) = 6$ and $f(4) = 6$. If we attempt to reverse this process, we have a situation where the input 6 would correspond to *two* potential outputs, 4 and 6.

However, the function g does have an inverse function because when we reverse the rows in Table 1.7.4, each input (in order, 3, 1, 4, 2, 0) indeed corresponds to one and only one output (in order, 0, 1, 2, 3, 4). We can thus make observations such as $g^{-1}(4) = 2$, which is the same as saying that $g(2) = 4$, just from a different perspective. □

In Example 1.7.2, we see that if we can identify one pair of distinct inputs that lead to the same output (such as $f(0) = f(4) = 6$ in Table 1.7.3), then the process of the function cannot be reversed and the function does not have an inverse.

Example 1.7.5 Do the functions p and q defined by Figure 1.7.6 and Figure 1.7.7 have corresponding inverse functions? Why or why not?

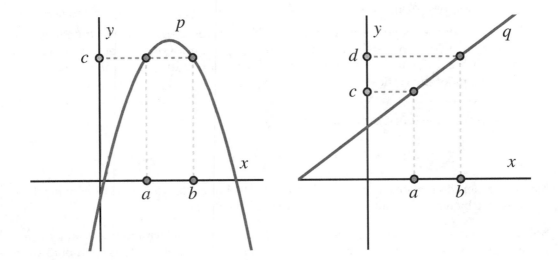

Figure 1.7.6: The graph that defines function p. **Figure 1.7.7:** The graph that defines function p.

Solution. Recall that when a point such as (a, c) lies on the graph of a function p, this means that the input $x = a$, which represents to a value on the horizontal axis, corresponds with the output $y = c$ that is represented by a value on the vertical axis. In this situation, we write $p(a) = c$. We note explicitly that p is a function because its graph passes the Vertical Line Test: any vertical line intersects the graph of p exactly, and thus each input from the domain corresponds to one and only one output.

If we attempt to change perspective and use the graph of p to view x as a function of y, we see that this fails because the output value c is associated with two different inputs, a and b. Said differently, because the horizontal line $y = c$ intersects the graph of p at both (a, c) and (b, c) (as shown in Figure 1.7.6), we cannot view y as the input to a function process that produces the corresonding x-value. Therefore, p does not have an inverse function.

On the other hand, provided that the behavior seen in the figure continues, the function q does have an inverse because we can view x as a function of y via the graph given in Figure 1.7.7. This is because for any choice of y, there corresponds one and only one x that results from y. We can think of this visually by starting at a value such as $y = c$ on the y-axis, moving horizontally to where the line intersects the graph of p, and then moving down to the corresonding location (here $x = a$) on the horizontal axis. From the behavior of the graph of q (a straight line that is always increasing), we see that this correspondence will hold for any choice of y, and thus indeed x is a function of y. From this, we can say that q indeed has an inverse function. We thus can write that $q^{-1}(c) = a$, which is a different way to express the equivalent fact that $q(a) = c$. □

The graphical observations that we made for the function q in Example 1.7.5 provide a general test for whether or not a function given by a graph has a corresponding inverse function.

Horizontal Line Test.

A function whose graph lies in the x-y plane has a corresponding inverse function if and only if every horizontal line intersects the graph at most once. When the graph passes this test, the horizontal coordinate of each point on the graph can be viewed as a function of the vertical coordinate of the point.

Example 1.7.8 Do the functions r and s defined by

$$y = r(t) = 3 - \frac{1}{5}(t - 1)^3 \text{ and } y = s(t) = 3 - \frac{1}{5}(t - 1)^2$$

have corresponding inverse functions? If not, use algebraic reasoning to explain why; if so, demonstrate by using algebra to find a formula for the inverse function.

Solution. For any function of the form $y = f(t)$, one way to determine if we can view the original input variable t as a function of the original output variable y is to attempt to solve the equation $y = f(t)$ for t in terms of y.

Taking $y = 3 - \frac{1}{5}(t - 1)^3$, we try to solve for t by first subtracting 3 from both sides to get

$$y - 3 = -\frac{1}{5}(t - 1)^3.$$

Next, multiplying both sides by -5, it follows that

$$(t - 1)^3 = -5(y - 3).$$

Because the cube root function has the property that $\sqrt[3]{z^3} = z$ for every real number z (since the cube root function is the inverse function for the cubing function, and each function has both a domain and range of all real numbers), we can take the cube root of both sides of the preceding equation to get

$$t - 1 = \sqrt[3]{-5(y - 3)}.$$

Finally, adding 1 to both sides, we have determined that

$$t = 1 + \sqrt[3]{-5(y - 3)}.$$

Because we have been able to express t as a single function of y for every possible value of y, this shows that r indeed has an inverse and that $t = r^{-1}(y) = 1 + \sqrt[3]{-5(y - 3)}$.

We attempt similar reasoning for the second function, $y = 3 - \frac{1}{5}(t - 1)^2$. To solve for t, we first subtract 3 from both sides, so that

$$y - 3 = -\frac{1}{5}(t - 1)^2.$$

After multiplying both sides by -5, we have

$$(t - 1)^2 = -5(y - 3).$$

Next, it's necessary to take the square root of both sides in an effort to isolate t. Here, however, we encounter a crucial issue. Because the function $g(x) = x^2$ takes any nonzero number and its opposite to the same output (e.g. $(-5)^2 = 25 = (5)^2$), this means that we have to account for *both* possible inputs that result in the same output. Based on our last equation, this means that either

$$t - 1 = \sqrt{-5(y - 3)} \text{ or } t - 1 = -\sqrt{-5(y - 3)}.$$

As such, we find not a single equation that expresses t as a function of y, but rather two:

$$t = 1 + \sqrt{-5(y - 3)} \text{ or } t = 1 - \sqrt{-5(y - 3)}.$$

Since it appears that t can't be expressed as a single function of y, it seems to follow that $y = s(t) = 3 - \frac{1}{5}(t - 1)^2$ does not have an inverse function. □

The graphs of $y = r(t) = 3 - \frac{1}{5}(t - 1)^3$ and $y = s(t) = 3 - \frac{1}{5}(t - 1)^2$ provide a different perspective to confirm the results of Example 1.7.8. Indeed, in Figure 1.7.9, we see that r appears to pass the horizontal line test because it is decreasing[2], and thus has an inverse function. On the other hand, the graph of s fails the horizontal line test (picture the line $y = 2$ in Figure 1.7.10) and therefore s does not have an inverse function.

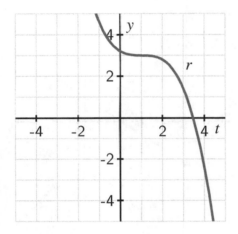 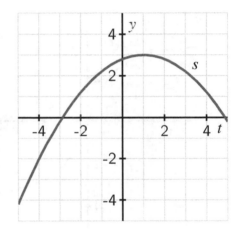

Figure 1.7.9: A plot of $y = r(t) = 3 - \frac{1}{5}(t - 1)^3$.

Figure 1.7.10: A plot of $y = s(t) = 3 - \frac{1}{5}(t - 1)^2$.

Activity 1.7.3. Determine, with justification, whether each of the following functions has an inverse function. For each function that has an inverse function, give two examples of values of the inverse function by writing statements such as "$s^{-1}(3) = 1$".

a. The function $f : S \to S$ given by Table 1.7.11, where $S = \{0, 1, 2, 3, 4\}$.

x	0	1	2	3	4
$f(x)$	1	2	4	3	2

Table 1.7.11: Values of $y = f(x)$.

[2]Calculus provides one way to fully justify that the graph of s is indeed always decreasing.

b. The function $g : S \rightarrow S$ given by Table 1.7.12, where $S = \{0, 1, 2, 3, 4\}$.

x	0	1	2	3	4
$f(x)$	4	0	3	1	2

Table 1.7.12: Values of $y = g(x)$.

c. The function p given by $p(t) = 7 - \frac{3}{5}t$. Assume that the domain and codomain of p are both "all real numbers".

d. The function q given by $q(t) = 7 - \frac{3}{5}t^4$. Assume that the domain and codomain of q are both "all real numbers".

e. The functions r and s given by the graphs in Figure 1.7.13 and Figure 1.7.14. Assume that the graphs show all of the important behavior of the functions and that the apparent trends continue beyond what is pictured.

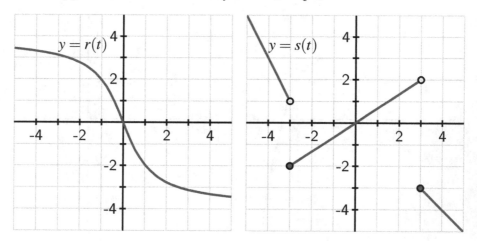

Figure 1.7.13: The graph of $y = r(t)$.   **Figure 1.7.14:** The graph of $y = s(t)$.

1.7.3 Properties of an inverse function

When a function has an inverse function, we have observed several important relationships that hold between the original function and the corresponding inverse function.

Properties of an inverse function.

Let $f : A \rightarrow B$ be a function whose domain is A and whose range is B be such that f has an inverse function, f^{-1}. Then:

- $f^{-1} : B \rightarrow A$, so the domain of f^{-1} is B and its range is A.

- The functions f and f^{-1} reverse one anothers' processes. Symbolically,

$f^{-1}(f(a)) = a$ for every input a in the domain of f, and similarly, $f(f^{-1}(b)) = b$ for every input b in the domain of f^{-1}.

- If $y = f(t)$, then we can express the exact same relationship from a different perspective by writing $t = f^{-1}(y)$.

- Consider the setting where A and B are collections of real numbers. If a point (x, y) lies on the graph of f, then it follows $y = f(x)$. From this, we can equivalently say that $x = f^{-1}(y)$. Hence, the point (y, x) lies on the graph of $x = f^{-1}(y)$.

The last item above leads to a special relationship between the graphs of f and f^{-1} when viewed on the same coordinate axes. In that setting, we need to view x as the input of each function (since it's the horizontal coordinate) and y as the output. If we know a particular input-output relationship for f, say $f(-1) = \frac{1}{2}$, then it follows that $f^{-1}\left(\frac{1}{2}\right) = -1$. We observe that the points $\left(-1, \frac{1}{2}\right)$ and $\left(\frac{1}{2}, -1\right)$ are reflections of each other across the line $y = x$. Because such a relationship holds for every point (x, y) on the graph of f, this means that the graphs of f and f^{-1} are reflections of one another across the line $y = x$, as seen in Figure 1.7.15.

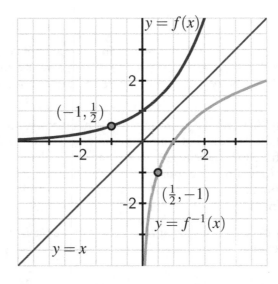

Figure 1.7.15: The graph of a function f along with its inverse, f^{-1}.

Activity 1.7.4. During a major rainstorm, the rainfall at Gerald R. Ford Airport is measured on a frequent basis for a 10-hour period of time. The following function g models the *rate*, R, at which the rain falls (in cm/hr) on the time interval $t = 0$ to $t = 10$:

$$R = g(t) = \frac{4}{t + 2} + 1$$

a. Compute $g(3)$ and write a complete sentence to explain its meaning in the given context, including units.

b. Compute the average rate of change of g on the time interval $[3, 5]$ and write two careful complete sentences to explain the meaning of this value in the context of the problem, including units. Explicitly address what the value you compute tells you about how rain is falling over a certain time interval, and what you should expect as time goes on.

c. Plot the function $y = g(t)$ using a computational device. On the domain $[1, 10]$, what is the corresponding range of g? Why does the function g have an inverse function?

d. Determine $g^{-1}\left(\frac{9}{5}\right)$ and write a complete sentence to explain its meaning in the given context.

e. According to the model g, is there ever a time during the storm that the rain falls at a rate of exactly 1 centimeter per hour? Why or why not? Provide an algebraic justification for your answer.

1.7.4 Summary

- A given function $f : A \to B$ has an inverse function whenever there exists a related function $g : B \to A$ that reverses the process of f. Formally, this means that g must satisfy $g(f(a)) = a$ for every a in the domain of f, and $f(g(b)) = b$ for every b in the range of f.

- We determine whether or not a given function f has a corresponding inverse function by determining if the process that defines f can be reversed so that we can also think of the outputs as a function of the inputs. If we have a graph of the function f, we know f has an inverse function if the graph passes the Horizontal Line Test. If we have a formula for the function f, say $y = f(t)$, we know f has an inverse function if we can solve for t and write $t = f^{-1}(y)$.

- A good summary of the properties of an inverse function is provided in the Properties of an inverse function.

1.7.5 Exercises

1. Suppose $P = f(t)$ is the population in millions in year t.

Which of the statements below best explains the meaning of the *INVERSE* function f^{-1}?

⊙ The population change over time

⊙ The year t in which the population is P million

⊙ The population P in millions in year t

⊙ How long it takes to reach P million

⊙ None of the above

2. Suppose $N = f(t)$ is the total number of inches of snow that fall in the first t days of January.

    Which of the statements below best explains the meaning of the **INVERSE** function f^{-1}?

    ⊙ The days for which there are N inches of snow on the ground

    ⊙ The number of days it takes to accumulate N inches of snow

    ⊙ The number of inches of snow accumulated in t days

    ⊙ The number of inches of snow on the ground after t days

    ⊙ None of the above

3. The cost (in dollars) of producing x air conditioners is $C = g(x) = 560 + 40x$. Find a formula for the inverse function $g^{-1}(C)$.

4. (a) Find a formula for the perimeter $P = f(s)$ of a square of side length s.

    (b) Find $f(4)$.

    Which of the statements best explains the meaning of $f(4)$?

    ⊙ The side length of a square of perimeter P

    ⊙ The area of a square of side length 4

    ⊙ The side length of a square of area 4

    ⊙ The perimeter of a square of side length 4

    (c) Find $f^{-1}(32)$.

    Which of the statements best explains the meaning of $f^{-1}(32)$?

    ⊙ The side length of a square of perimeter 32

    ⊙ The area of a square of side length 32

    ⊙ The side length of a square of area 32

    ⊙ The perimeter of a square of side length 32

    (d) Find a formula for the inverse function $f^{-1}(P)$.

5. Suppose $V = f(t)$ is the speed in km/hr of an accelerating car t seconds after starting.

    Which of the statements best explains the meaning of the **INVERSE** function f^{-1}?

     ⊙ The acceleration of a car which is going V km/hr.

     ⊙ The number of seconds it takes a car to reach a speed of V km/hr.

     ⊙ How long after leaving the car has an acceleration of V km per second squared.

     ⊙ The velocity of a car t seconds after accelerating.

     ⊙ None of the above

6. The gross domestic product (GDP) of the US is given by $G(t)$ where t is the number of years since 1990, and the units of G are billions of dollars. Match the meaning of each of the mathematical expressions below with the correct description below.

    (a) $G^{-1}(9873)$

    (b) $G(11)$

    A. How many years after 1990 it was when the GDP was 9,873 dollars.

    B. How many billions of dollars the GDP was in 2001.

    C. The year the GDP was 9,873 billion dollars.

    D. How many dollars the GDP is expected to be in 11 years.

    E. How many billions of dollars the GDP was in 1991.

    F. How many years after 1990 it was when the GDP was 9,873 billion dollars.

7. Consider the functions p and q whose graphs are given by Figure 1.7.16

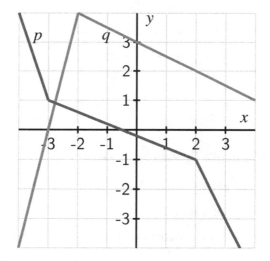

Figure 1.7.16: Plots of the graphs of p and q.

    a. Compute each of the following values exactly, or explain why they are not defined: $p^{-1}(2.5)$, $p^{-1}(-2)$, $p^{-1}(0)$, and $q^{-1}(2)$.

b. From your work in (a), you know that the point $(2.5, -3.5)$ lies on the graph of p^{-1}. In addition to the other two points you know from (a), find three additional points that lie on the graph of p^{-1}.

c. On Figure 1.7.16, plot the 6 points you have determined in (a) and (b) that lie on the graph of $y = p^{-1}(x)$. Then, sketch the complete graph of $y = p^{-1}(x)$. How are the graphs of p and p^{-1} related to each other?

8. Consider an inverted conical tank that is being filled with water. The tank's radius is 2 m and its depth is 4 m. Suppose the tank is initially empty and is being filled in such a way that the height of the water is always rising at a rate of 0.25 meters per minute.

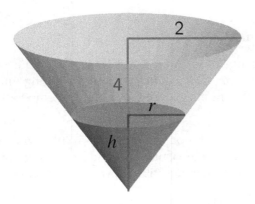

Figure 1.7.17: The conical tank.

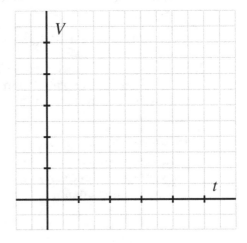

Figure 1.7.18: Axes to plot $V = g(t)$.

a. Explain why the height, h, of the water can be viewed as a function of t according to the formula $h = f(t) = 0.25t$.

b. At what time is the water in the tank 2.5 m deep? At what time is the tank completely full?

c. Suppose we think of the volume, V, of water in the tank as a function of t and name the function $V = g(t)$. Do you expect that the function g has an inverse function? Why or why not?

d. Recall that the volume of a cone of radius r and height h is $V = \frac{\pi}{3}r^2 h$. Due to the shape of the tank, similar triangles tell us that r and h satisfy the proportion $r = \frac{1}{2}h$, and thus

$$V = \frac{\pi}{3}\left(\frac{1}{2}h\right)^2 h = \frac{\pi}{12}h^3. \tag{1.7.1}$$

Use the fact that $h = f(t) = 0.25t$ along with Equation (1.7.1) to find a formula for $V = g(t)$. Sketch a plot of $V = g(t)$ on the blank axes provided in Figure 1.7.18. Write at least one sentence to explain why $V = g(t)$ has the shape that it does.

   e. Take the formula for $V = g(t)$ that you determined in (d) and solve for t to determine a formula for $t = g^{-1}(V)$. What is the meaning of the formula you find?

   f. Find the exact time that there is $\frac{8}{3}\pi$ cubic meters of volume in the tank.

9.   Recall that in Activity 1.6.3, we showed that Celsius temperature is a function of the number of chirps per minute from a snowy tree cricket according to the formula

$$C = H(N) = \frac{40}{9} + \frac{5}{36}N. \hspace{2cm} (1.7.2)$$

   a. What familiar type of function is H? Why must H have an inverse function?

   b. Determine an algebraic formula for $N = H^{-1}(C)$. Clearly show your work and thinking.

   c. What is the meaning of the statement $72 = H^{-1}\left(\frac{130}{9}\right)$?

   d. Determine the average rate of change of H on the interval $[40, 50]$. Write a complete sentence to explain the meaning of the value you find, including units on the value. Explain clearly how this number describes how the temperature is changing.

   e. Determine the average rate of change of H^{-1} on the interval $[15, 20]$. Write a complete sentence to explain the meaning of the value you find, including units on the value. Explain clearly how this number describes how the number of chirps per minute is changing.

1.8 Transformations of Functions

Motivating Questions

- How is the graph of $y = g(x) = af(x - b) + c$ related to the graph of $y = f(x)$?

- What do we mean by "transformations" of a given function f? How are translations and vertical stretches of a function examples of transformations?

In our preparation for calculus, we aspire to understand functions from a wide range of perspectives and to become familiar with a library of basic functions. So far, two basic families functions we have considered are linear functions and quadratic functions, the simplest of which are $L(x) = x$ and $Q(x) = x^2$. As we progress further, we will endeavor to understand a "parent" function as the most fundamental member of a family of functions, as well as how other similar but more complicated functions are the result of transforming the parent function.

Informally, a transformation of a given function is an algebraic process by which we change the function to a related function that has the same fundamental shape, but may be shifted, reflected, and/or stretched in a systematic way. For example, among all quadratic functions, the simplest is the parent function $Q(x) = x^2$, but any other quadratic function such as $g(x) = -3(x - 5)^2 + 4$ can also be understood in relation to the parent function. We say that "g is a transformation of f."

In Preview Activity 1.8.1, we investigate the effects of the constants a, b, and c in generating the function $g(x) = af(x - b) + c$ in the context of already knowing the function f.

Preview Activity 1.8.1. Open a new *Desmos* graph and define the function $f(x) = x^2$. Adjust the window so that the range is for $-4 \leq x \leq 4$ and $-10 \leq y \leq 10$.

a. In *Desmos*, define the function $g(x) = f(x) + a$. (That is, in *Desmos* on line 2, enter g(x) = f(x) + a.) You will get prompted to add a slider for a. Do so.

Explore by moving the slider for a and write at least one sentence to describe the effect that changing the value of a has on the graph of g.

b. Next, define the function $h(x) = f(x - b)$. (That is, in *Desmos* on line 4, enter h(x) = f(x-b) and add the slider for b.)

Move the slider for b and write at least one sentence to describe the effect that changing the value of b has on the graph of h.

c. Now define the function $p(x) = cf(x)$. (That is, in *Desmos* on line 6, enter p(x) = cf(x) and add the slider for c.)

Move the slider for c and write at least one sentence to describe the effect that changing the value of c has on the graph of p. In particular, when $c = -1$, how is the graph of p related to the graph of f?

d. Finally, click on the icons next to g, h, and p to temporarily hide them, and go back to Line 1 and change your formula for f. You can make it whatever

> you'd like, but try something like $f(x) = x^2 + 2x + 3$ or $f(x) = x^3 - 1$. Then, investigate with the sliders a, b, and c to see the effects on g, h, and p (unhiding them appropriately). Write a couple of sentences to describe your observations of your explorations.

1.8.1 Translations of Functions

We begin by summarizing two of our findings in Preview Activity 1.8.1.

Vertical Translation of a Function.

Given a function $y = f(x)$ and a real number a, the transformed function $y = g(x) = f(x) + a$ is a *vertical translation* of the graph of f. That is, every point $(x, f(x))$ on the graph of f gets shifted vertically to the corresponding point $(x, f(x)+a)$ on the graph of g.

As we found in our *Desmos* explorations in the preview activity, is especially helpful to see the effects of vertical translation dynamically.

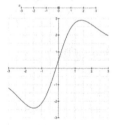   

Figure 1.8.1: Interactive vertical translations demonstration (in the HTML version only).

In a vertical translation, the graph of g lies above the graph of f whenever $a > 0$, while the graph of g lies below the graph of f whenever $a < 0$. In Figure 1.8.2, we see the original parent function $f(x) = |x|$ along with the resulting transformation $g(x) = f(x) - 3$, which is a downward vertical shift of 3 units. Note particularly that every point on the original graph of f is moved 3 units down; we often indicate this by an arrow and labeling at least one key point on each graph.

In Figure 1.8.3, we see a horizontal translation of the original function f that shifts its graph 2 units to the right to form the function h. Observe that f is not a familiar basic function; transformations may be applied to any original function we desire.

From an algebraic point of view, horizontal translations are slightly more complicated than vertical ones. Given $y = f(x)$, if we define the transformed function $y = h(x) = f(x - b)$, observe that

$$h(x + b) = f((x + b) - b) = f(x).$$

---

[1]Huge thanks to the amazing David Austin for making these interactive javascript graphics for the text.

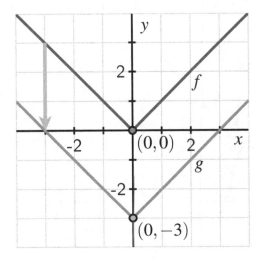
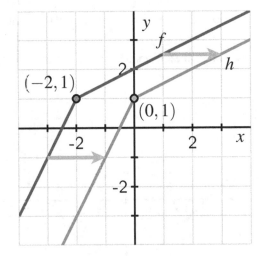

Figure 1.8.2: A vertical translation, g, of the function $y = f(x) = |x|$.

Figure 1.8.3: A horizontal translation, h, of a different function $y = f(x)$.

This shows that for an input of $x + b$ in h, the output of h is the same as the output of f that corresponds to an input of simply x. Hence, in Figure 1.8.3, the formula for h in terms of f is $h(x) = f(x - 2)$, since an input of $x + 2$ in h will result in the same output as an input of x in f. For example, $h(2) = f(0)$, which aligns with the graph of h being a shift of the graph of f to the right by 2 units.

Again, it's instructive to see the effects of horizontal translation dynamically.

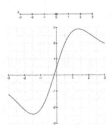

Figure 1.8.4: Interactive horizontal translations demonstration (in the HTML version only).

Overall, we have the following general principle.

Horizontal Translation of a Function.

Given a function $y = f(x)$ and a real number b, the transformed function $y = h(x) = f(x - b)$ is a *horizontal translation* of the graph of f. That is, every point $(x, f(x))$ on the graph of f gets shifted horizontally to the corresponding point $(x + b, f(x))$ on the graph of g.

We emphasize that in the horizontal translation $h(x) = f(x - b)$, if $b > 0$ the graph of h lies b units to the right of f, while if $b < 0$, h lies b units to the left of f.

Activity 1.8.2. Consider the functions r and s given in Figure 1.8.5 and Figure 1.8.6.

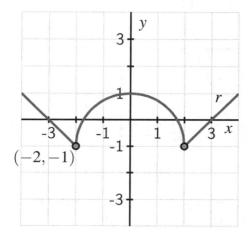

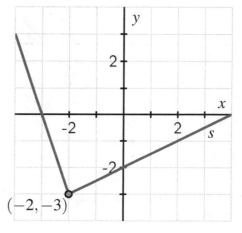

Figure 1.8.5: A parent function r.   **Figure 1.8.6:** A parent function s.

a. On the same axes as the plot of $y = r(x)$, sketch the following graphs: $y = g(x) = r(x) + 2$, $y = h(x) = r(x + 1)$, and $y = f(x) = r(x + 1) + 2$. Be sure to label the point on each of g, h, and f that corresponds to $(-2, -1)$ on the original graph of r. In addition, write one sentence to explain the overall transformations that have resulted in g, h, and f.

b. Is it possible to view the function f in (a) as the result of composition of g and h? If so, in what order should g and h be composed in order to produce f?

c. On the same axes as the plot of $y = s(x)$, sketch the following graphs: $y = k(x) = s(x) - 1$, $y = j(x) = s(x - 2)$, and $y = m(x) = s(x - 2) - 1$. Be sure to label the point on each of k, j, and m that corresponds to $(-2, -3)$ on the original graph of r. In addition, write one sentence to explain the overall transformations that have resulted in k, j, and m.

d. Now consider the function $q(x) = x^2$. Determine a formula for the function that is given by $p(x) = q(x + 3) - 4$. How is p a transformation of q?

1.8.2 Vertical stretches and reflections

So far, we have seen the possible effects of adding a constant value to function output —$f(x)+a$— and adding a constant value to function input — $f(x+b)$. Each of these actions results in a translation of the function's graph (either vertically or horizontally), but otherwise leaving the graph the same. Next, we investigate the effects of multiplication the function's output by a constant.

Example 1.8.7 Given the parent function $y = f(x)$ pictured in Figure 1.8.8, what are the effects of the transformation $y = v(x) = cf(x)$ for various values of c?

Solution. We first investigate the effects of $c = 2$ and $c = \frac{1}{2}$. For $v(x) = 2f(x)$, the algebraic impact of this transformation is that every output of f is multiplied by 2. This means that the only output that is unchanged is when $f(x) = 0$, while any other point on the graph of the original function f will be stretched vertically away from the x-axis by a factor of 2. We can see this in Figure 1.8.8 where each point on the original dark blue graph is transformed to a corresponding point whose y-coordinate is twice as large, as partially indicated by the red arrows.

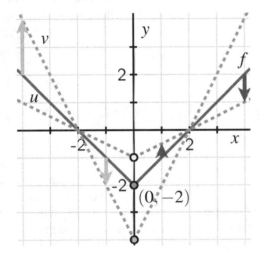
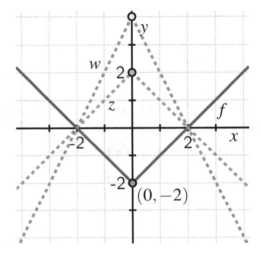

Figure 1.8.8: The parent function $y = f(x)$ along with two different vertical stretches, v and u.

Figure 1.8.9: The parent function $y = f(x)$ along with a vertical reflection, z, and a corresponding stretch, w.

In contrast, the transformation $u(x) = \frac{1}{2}f(x)$ is stretched vertically by a factor of $\frac{1}{2}$, which has the effect of compressing the graph of f towards the x-axis, as all function outputs of f are multiplied by $\frac{1}{2}$. For instance, the point $(0, -2)$ on the graph of f is transformed to the graph of $(0, -1)$ on the graph of u, and others are transformed as indicated by the purple arrows.

To consider the situation where $c < 0$, we first consider the simplest case where $c = -1$ in the transformation $z(x) = -f(x)$. Here the impact of the transformation is to multiply every output of the parent function f by -1; this takes any point of form (x, y) and transforms it to $(x, -y)$, which means we are reflecting each point on the original function's graph across the x-axis to generate the resulting function's graph. This is demonstrated in Figure 1.8.9 where $y = z(x)$ is the reflection of $y = f(x)$ across the x-axis.

Finally, we also investigate the case where $c = -2$, which generates $y = w(x) = -2f(x)$. Here we can think of -2 as $-2 = 2(-1)$: the effect of multiplying by -1 first reflects the graph of f across the x-axis (resulting in w), and then multiplying by 2 stretches the graph of z vertically to result in w, as shown in Figure 1.8.9. □

As with vertical and horizontal translation, it's particularly instructive to see the effects of vertical scaling in a dynamic way.

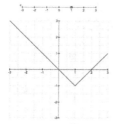

Figure 1.8.10: Interactive vertical scaling demonstration (in the HTML version only).

We summarize and generalize our observations from Example 1.8.7 and Figure 1.8.10 as follows.

> **Vertical Scaling of a Function.**
>
> Given a function $y = f(x)$ and a real number $c > 0$, the transformed function $y = v(x) = cf(x)$ is a *vertical stretch* of the graph of f. Every point $(x, f(x))$ on the graph of f gets stretched vertically to the corresponding point $(x, cf(x))$ on the graph of v. If $0 < c < 1$, the graph of v is a compression of f toward the x-axis; if $c > 1$, the graph of v is a stretch of f away from the x-axis. Points where $f(x) = 0$ are unchanged by the transformation.
>
> Given a function $y = f(x)$ and a real number $c < 0$, the transformed function $y = v(x) = cf(x)$ is a reflection of the graph of f across the x-axis followed by a vertical stretch by a factor of $|c|$.

Activity 1.8.3. Consider the functions r and s given in Figure 1.8.11 and Figure 1.8.12.

  a. On the same axes as the plot of $y = r(x)$, sketch the following graphs: $y = g(x) = 3r(x)$ and $y = h(x) = \frac{1}{3}r(x)$. Be sure to label several points on each of r, g, and h with arrows to indicate their correspondence. In addition, write one sentence to explain the overall transformations that have resulted in g and h from r.

  b. On the same axes as the plot of $y = s(x)$, sketch the following graphs: $y = k(x) = -s(x)$ and $y = j(x) = -\frac{1}{2}s(x)$. Be sure to label several points on each of s, k, and j with arrows to indicate their correspondence. In addition, write one sentence to explain the overall transformations that have resulted in k and j from s.

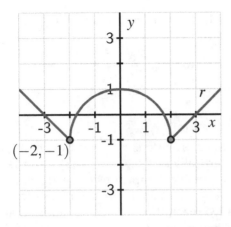

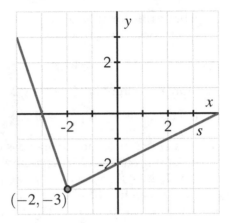

Figure 1.8.11: A parent function r.   **Figure 1.8.12:** A parent function s.

c. On the additional copies of the two figures below, sketch the graphs of the following transformed functions: $y = m(x) = 2r(x + 1) - 1$ (at left) and $y = n(x) = \frac{1}{2}s(x-2)+2$. As above, be sure to label several points on each graph and indicate their correspondence to points on the original parent function.

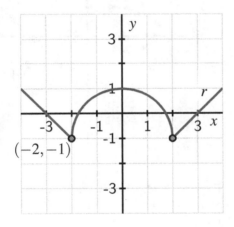

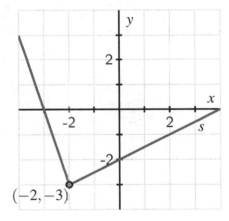

d. Describe in words how the function $y = m(x) = 2r(x + 1) - 1$ is the result of three elementary transformations of $y = r(x)$. Does the order in which these transformations occur matter? Why or why not?

1.8.3 Combining shifts and stretches: why order sometimes matters

In the final question of Activity 1.8.3, we considered the transformation $y = m(x) = 2r(x + 1) - 1$ of the original function r. There are three different basic transformations involved: a vertical shift of 1 unit down, a horizontal shift of 1 unit left, and a vertical stretch by a factor of 2. To understand the order in which these transformations are applied, it's essential to remember that a function is a *process* that converts inputs to outputs.

By the algebraic rule for m, $m(x) = 2r(x + 1) - 1$. In words, this means that given an input x for m, we do the following processes in this particular order:

1. add 1 to x and then apply the function r to the quantity $x + 1$;

2. multiply the output of $r(x + 1)$ by 2;

3. subtract 1 from the output of $2r(x + 1)$.

These three steps correspond to three basic transformations: (1) shift the graph of r to the left by 1 unit; (2) stretch the resulting graph vertically by a factor of 2; (3) shift the resulting graph vertically by -1 units. We can see the graphical impact of these algebraic steps by taking them one at a time. In Figure 1.8.14, we see the function p that results from a shift 1 unit left of the parent function in Figure 1.8.13. (Each time we take an additional step, we will de-emphasize the preceding function by having it appear in lighter color and dashed.)

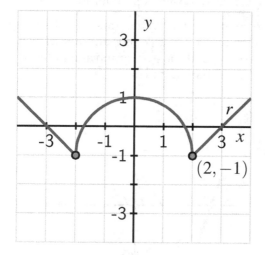 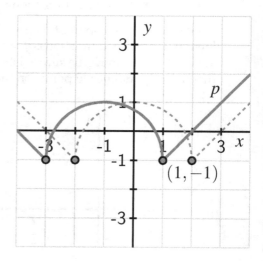

Figure 1.8.13: The parent function $y = r(x)$. **Figure 1.8.14:** The parent function $y = r(x)$ along with the horizontal shift $y = p(x) = r(x + 1)$.

Continuing, we now consider the function $q(x) = 2p(x) = 2r(x+1)$, which results in a vertical stretch of p away from the x-axis by a factor of 2, as seen in Figure 1.8.15.

Finally, we arrive at $y = m(x) = 2r(x + 1) - 1$ by subtracting 1 from $q(x) = 2r(x + 1)$; this of course is a vertical shift of -1 units, and produces the graph of m shown in red in Figure 1.8.16. We can also track the point $(2, -1)$ on the original parent function: it first moves left 1 unit to $(1, -1)$, then it is stretched vertically by a factor of 2 away from the x-axis to $(1, -2)$, and lastly is shifted 1 unit down to the point $(1, -3)$, which we see on the graph of m.

While there are some transformations that can be executed in either order (such as a combination of a horizontal translation and a vertical translation, as seen in part (b) of Activity 1.8.2), in other situations order matters. For instance, in our preceding discussion, we

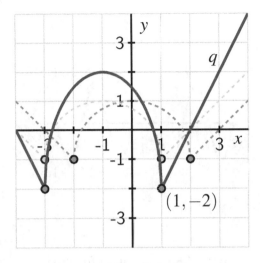

Figure 1.8.15: The function $y = q(x) = 2p(x) = 2r(x + 1)$ along with graphs of p and r.

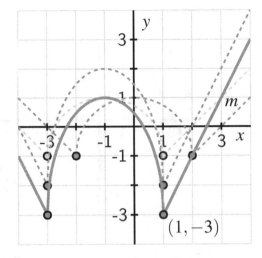

Figure 1.8.16: The function $y = m(x) = q(x) - 1 = 2r(x + 1) - 1$ along with graphs of q, p and r.

have to apply the vertical stretch *before* applying the vertical shift. Algebraically, this is because

$$2r(x + 1) - 1 \neq 2[r(x + 1) - 1].$$

The quantity $2r(x + 1) - 1$ multiplies the function $r(x + 1)$ by 2 first (the stretch) and then the vertical shift follows; the quantity $2[r(x + 1) - 1]$ shifts the function $r(x + 1)$ down 1 unit first, and then executes a vertical stretch by a factor of 2. In the latter scenario, the point $(1, -1)$ that lies on $r(x + 1)$ gets transformed first to $(1, -2)$ and then to $(1, -4)$, which is not the same as the point $(1, -3)$ that lies on $m(x) = 2r(x + 1) - 1$.

Activity 1.8.4. Consider the functions f and g given in Figure 1.8.17 and Figure 1.8.18.

a. Sketch an accurate graph of the transformation $y = p(x) = -\frac{1}{2}f(x - 1) + 2$. Write at least one sentence to explain how you developed the graph of p, and identify the point on p that corresponds to the original point $(-2, 2)$ on the graph of f.

b. Sketch an accurate graph of the transformation $y = q(x) = 2g(x + 0.5) - 0.75$. Write at least one sentence to explain how you developed the graph of p, and identify the point on q that corresponds to the original point $(1.5, 1.5)$ on the graph of g.

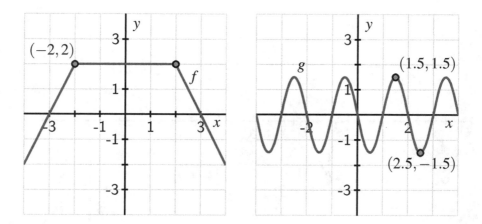

Figure 1.8.17: A parent function f.   **Figure 1.8.18:** A parent function g.

c. Is the function $y = r(x) = \frac{1}{2}(-f(x-1) - 4)$ the same function as p or different? Why? Explain in two different ways: discuss the algebraic similarities and differences between p and r, and also discuss how each is a transformation of f.

d. Find a formula for a function $y = s(x)$ (in terms of g) that represents this transformation of g: a horizontal shift of 1.25 units left, followed by a reflection across the x-axis and a vertical stretch of 2.5 units, followed by a vertical shift of 1.75 units. Sketch an accurate, labeled graph of s on the following axes along with the given parent function g.

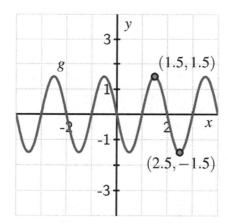

1.8.4 Summary

- The graph of $y = g(x) = af(x - b) + c$ is related to the graph of $y = f(x)$ by a sequence of transformations. First, there is horizontal shift of $|b|$ units to the right ($b > 0$) or left ($b < 0$). Next, there is a vertical stretch by a factor of $|a|$ (along with a reflection across $y = 0$ in the case where $a < 0$). Finally, there's a vertical shift of c units.

- A transformation of a given function f is a process by which the graph may be shifted or stretched to generate a new, related function with fundamentally the same shape. In this section we considered four different ways this can occur: through a horizontal translation (shift), through a reflection across the line $y = 0$ (the x-axis), through a vertical scaling (stretch) that multiplies every output of a function by the same constant, and through a vertical translation (shift). Each of these individual processes is itself a transformation, and they may be combined in various ways to create more complicated transformations.

1.8.5 Exercises

1. The graph of $y = x^2$ is given below:

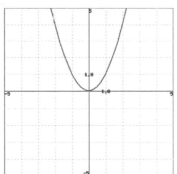

Find a formula for each of the transformations whose graphs are given below.

a)

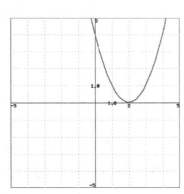

b)

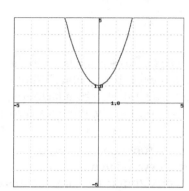

2. To obtain a new graph, stretch the graph of a function $f(x)$ vertically by a factor of 6. Then shift the new graph 4 units to the right and 2 units up. The result is the graph of a function

$$g(x) = Af(x + B) + C$$

where A, B, C are certain numbers. What are A, B, and C?

3. Identify the graphs A (blue), B (red) and C (green):

_____ is the graph of the function $f(x) = (x - 6)^2$

_____ is the graph of the function $g(x) = (x + 3)^2$

_____ is the graph of the function $h(x) = x^2 - 5$

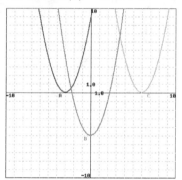

4.

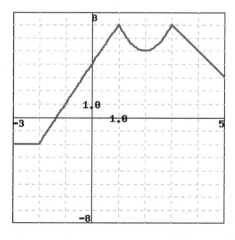

The figure above is the graph of the function $m(t)$. Let $n(t) = m(t) + 2$, $k(t) = m(t + 1.5)$, $w(t) = m(t - 0.5) - 2.5$ and $p(t) = m(t - 1)$. Find the values of the following:

1. $n(-3)$

2. $n(1)$

3. $k(2)$

4. $w(1.5)$

5. $w(-1.5)$

6. $p(2)$

5.   The graph of $f(x)$ contains the point $(9, 4)$. What point must be on each of the following transformed graphs?

    (a) The graph of $f(x - 6)$ must contain the point _____

    (b) The graph of $f(x) - 5$ must contain the point _____

    (c) The graph of $f(x + 2) + 7$ must contain the point _____

6.   Let $f(x) = x^2$.

    a. Let $g(x) = f(x) + 5$. Determine $AV_{[-3,-1]}$ and $AV_{[2,5]}$ for both f and g. What do you observe? Why does this phenomenon occur?

    b. Let $h(x) = f(x - 2)$. For f, recall that you determined $AV_{[-3,-1]}$ and $AV_{[2,5]}$ in (a). In addition, determine $AV_{[-1,-1]}$ and $AV_{[4,7]}$ for h. What do you observe? Why does this phenomenon occur?

    c. Let $k(x) = 3f(x)$. Determine $AV_{[-3,-1]}$ and $AV_{[2,5]}$ for k, and compare the results to your earlier computations of $AV_{[-3,-1]}$ and $AV_{[2,5]}$ for f. What do you observe? Why does this phenomenon occur?

    d. Finally, let $m(x) = 3f(x - 2) + 5$. Without doing any computations, what do you think will be true about the relationship between $AV_{[-3,-1]}$ for f and $AV_{[-1,1]}$ for m? Why? After making your conjecture, execute appropriate computations to see if your intuition is correct.

7.   Consider the parent function $y = f(x) = x$.

    a. Consider the linear function in point-slope form given by $y = L(x) = -4(x-3)+5$. What is the slope of this line? What is the most obvious point that lies on the line?

    b. How can the function L given in (a) be viewed as a transformation of the parent function f? Explain the roles of 3, -4, and 5, respectively.

    c. Explain why any non-vertical line of the form $P(x) = m(x-x_0)+y_0$ can be thought of as a transformation of the parent function $f(x) = x$. Specifically discuss the transformation(s) involved.

    d. Find a formula for the transformation of $f(x) = x$ that corresponds to a horizontal shift of 7 units left, a reflection across $y = 0$ and vertical stretch of 3 units away from the x-axis, and a vertical shift of -11 units.

8. We have explored the effects of adding a constant to the output of a function, $y = f(x) + a$, adding a constant to the input, $y = f(x + a)$, and multiplying the output of a function by a constant, $y = af(x)$. There is one remaining natural transformation to explore: multiplying the input to a function by a constant. In this exercise, we consider the effects of the constant a in transforming a parent function f by the rule $y = f(ax)$.

    Let $f(x) = (x - 2)^2 + 1$.

    a. Let $g(x) = f(4x)$, $h(x) = f(2x)$, $k(x) = f(0.5x)$, and $m(x) = f(0.25x)$. Use *Desmos* to plot these functions. Then, sketch and label g, h, k, and m on the provided axes in Figure 1.8.19 along with the graph of f. For each of the functions, label and identify its vertex, its y-intercept, and its x-intercepts.

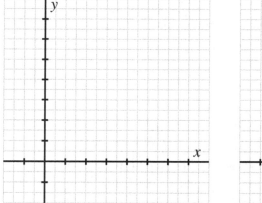 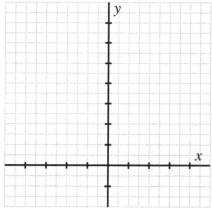

Figure 1.8.19: Axes for plotting f, g, h, k, and m in part (a). **Figure 1.8.20:** Axes for plotting f, r, and s from parts (c) and (d).

    b. Based on your work in (a), how would you describe the effect(s) of the transformation $y = f(ax)$ where $a > 0$? What is the impact on the graph of f? Are any parts of the graph of f unchanged?

    c. Now consider the function $r(x) = f(-x)$. Observe that $r(-1) = f(1)$, $r(2) = f(-2)$, and so on. Without using a graphing utility, how do you expect the graph of $y = r(x)$ to compare to the graph of $y = f(x)$? Explain. Then test your conjecture by using a graphing utility and record the plots of f and r on the axes in Figure 1.8.20.

    d. How do you expect the graph of $s(x) = f(-2x)$ to appear? Why? More generally, how does the graph of $y = f(ax)$ compare to the graph of $y = f(x)$ in the situation where $a < 0$?

1.9 Combining Functions

Motivating Questions

- How can we create new functions by adding, subtracting, multiplying, or dividing given functions?

- What are piecewise functions and what are different ways we can represent them?

In arithmetic, we execute processes where we take two numbers to generate a new number. For example, $2 + 3 = 5$: the number 5 results from adding 2 and 3. Similarly, we can multiply two numbers to generate a new one: $2 \cdot 3 = 6$.

We can work similarly with functions. Indeed, we have already seen a sophisticated way to combine two functions to generate a new, related function through composition. If $g : A \to B$ and $f : B \to C$, then we know there's a new, related function $f \circ g : A \to C$ defined by the process $(f \circ g)(x) = f(g(x))$. Said differently, the new function $f \circ g$ results from executing g first, followed by f.

Just as we can add, subtract, multiply, and divide numbers, we can also add, subtract, multiply, and divide functions to create a new function from two or more given functions.

Preview Activity 1.9.1. Consider the functions f and g defined by Table 1.9.1 and functions p and q defined by Figure 1.9.2.

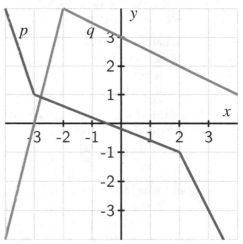

x	0	1	2	3	4
$f(x)$	5	10	15	20	25
$g(x)$	9	5	3	2	3

Table 1.9.1: Table defining functions f and g.

Figure 1.9.2: Graphs defining functions p and q.

a. Let $h(x) = f(x) + g(x)$. Determine $h(3)$.

b. Let $r(x) = p(x) - q(x)$. Determine $r(-1)$ exactly.

    c. Are there any values of x for which $r(x) = 0$? If not, explain why; if so, determine all such values, with justification.

    d. Let $k(x) = f(x) \cdot g(x)$. Determine $k(0)$.

    e. Let $s(x) = \frac{p(x)}{q(x)}$. Determine $s(1)$ exactly.

    f. Are there any values of x in the interval $-4 \le x \le 4$ for which $s(x)$ is not defined? If not, explain why; if so, determine all such values, with justification.

1.9.1 Arithmetic with functions

In most mathematics up until calculus, the main object we study is *numbers*. We ask questions such as

- "what number(s) form solutions to the equation $x^2 - 4x - 5 = 0$?"

- "what number is the slope of the line $3x - 4y = 7$?"

- "what number is generated as output by the function $f(x) = \sqrt{x^2 + 1}$ by the input $x = -2$?"

Certainly we also study overall patterns as seen in functions and equations, but this usually occurs through an examination of numbers themselves, and we think of numbers as the main objects being acted upon.

This changes in calculus. In calculus, the fundamental objects being studied are functions themselves. A function is a much more sophisticated mathematical object than a number, in part because a function can be thought of in terms of its graph, which is an infinite collection of ordered pairs of the form $(x, f(x))$.

It is often helpful to look at a function's formula and observe algebraic structure. For instance, given the quadratic function

$$q(x) = -3x^2 + 5x - 7$$

we might benefit from thinking of this as the sum of three simpler functions: the constant function $c(x) = -7$, the linear function $s(x) = 5x$ that passes through $(0, 0)$ with slope $m = 5$, and the concave down basic quadratic function $w(x) = -3x^2$. Indeed, each of the simpler functions c, s, and w contribute to making q be the function that it is. Likewise, if we were interested in the function $p(x) = (3x^2 + 4)(9 - 2x^2)$, it might be natural to think about the two simpler functions $f(x) = 3x^2 + 4$ and $g(x) = 9 - 2x^2$ that are being multiplied to produce p.

We thus naturally arrive at the ideas of adding, subtracting, multiplying, or dividing two or more functions, and hence introduce the following definitions and notation.

Definition 1.9.3 Let f and g be functions that share the same domain. Then,

- The **sum of f and g** is the function $f + g$ defined by $(f + g)(x) = f(x) + g(x)$.

- The **difference of** f **and** g is the function $f - g$ defined by $(f - g)(x) = f(x) - g(x)$.

- The **product of** f **and** g is the function $f \cdot g$ defined by $(f \cdot g)(x) = f(x) \cdot g(x)$.

- The **quotient of** f **and** g is the function $\frac{f}{g}$ defined by $\left(\frac{f}{g}\right)(x) = \frac{f(x)}{g(x)}$ for all x such that $g(x) \neq 0$.

◊

Activity 1.9.2. Consider the functions f and g defined by Figure 1.9.4 and Figure 1.9.5.

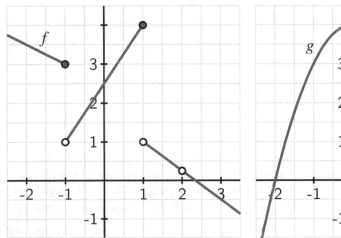

Figure 1.9.4: The function f.

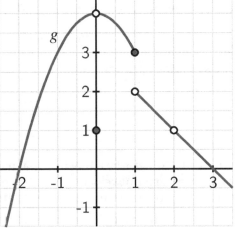

Figure 1.9.5: The function g.

a. Determine the exact value of $(f + g)(0)$.

b. Determine the exact value of $(g - f)(1)$.

c. Determine the exact value of $(f \cdot g)(-1)$.

d. Are there any values of x for which $\left(\frac{f}{g}\right)(x)$ is undefined? If not, explain why. If so, determine the values and justify your answer.

e. For what values of x is $(f \cdot g)(x) = 0$? Why?

f. Are there any values of x for which $(f - g)(x) = 0$? Why or why not?

1.9.2 Combining functions in context

When we work in applied settings with functions that model phenomena in the world around us, it is often useful to think carefully about the units of various quantities. Analyzing units can help us both understand the algebraic structure of functions and the variables involved, as well as assist us in assigning meaning to quantities we compute. We have already seen this with the notion of average rate of change: if a function $P(t)$ measures the population in a city in year t and we compute $AV_{[5,11]}$, then the units on $AV_{[5,11]}$ are "people per year," and the value of $AV_{[5,11]}$ is telling us the average rate at which the population changes in people per year on the time interval from year 5 to year 11.

Example 1.9.6 Say that an investor is regularly purchasing stock in a particular company.[1] Let $N(t)$ represent the number of shares owned on day t, where $t = 0$ represents the first day on which shares were purchased. Let $S(t)$ give the value of one share of the stock on day t; note that the units on $S(t)$ are dollars per share. How is the total value, $V(t)$, of the held stock on day t determined?

Solution. Observe that the units on $N(t)$ are "shares" and the units on $S(t)$ are "dollars per share". Thus when we compute the product

$$N(t)\,\text{shares} \cdot S(t)\,\text{dollars per share,}$$

it follows that the resulting units are "dollars", which is the total value of held stock. Hence,

$$V(t) = N(t) \cdot S(t).$$

□

Activity 1.9.3. Let f be a function that measures a car's fuel economy in the following way. Given an input velocity v in miles per hour, $f(v)$ is the number of gallons of fuel that the car consumes per mile (i.e., "gallons per mile"). We know that $f(60) = 0.04$.

a. What is the meaning of the statement "$f(60) = 0.04$" in the context of the problem? That is, what does this say about the car's fuel economy? Write a complete sentence.

b. Consider the function $g(v) = \frac{1}{f(v)}$. What is the value of $g(60)$? What are the units on g? What does g measure?

c. Consider the function $h(v) = v \cdot f(v)$. What is the value of $h(60)$? What are the units on h? What does h measure?

d. Do $f(60)$, $g(60)$, and $h(60)$ tell us fundamentally different information, or are they all essentially saying the same thing? Explain.

e. Suppose we also know that $f(70) = 0.045$. Find the average rate of change of f on the interval $[60,70]$. What are the units on the average rate of change of f? What does this quantity measure? Write a complete sentence to explain.

1.9.3 Piecewise functions

In both abstract and applied settings, we sometimes have to use different formulas on different intervals in order to define a function of interest.

---

[1]This example is taken from Section 2.3 of Active Calculus.

A familiar and important function that is defined *piecewise* is the absolute value function: $A(x) = |x|$. We know that if $x \geq 0$, $|x| = x$, whereas if $x < 0$, $|x| = -x$.

Definition 1.9.7 The absolute value of a real number, denoted by $A(x) = |x|$, is defined by the rule

$$A(x) = \begin{cases} -x, & x < 0 \\ x, & x \geq 0 \end{cases}$$

◊

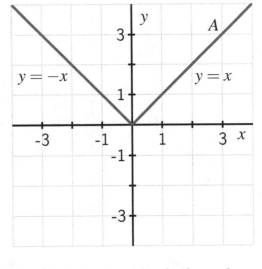

Figure 1.9.8: A plot of the absolute value function, $A(x) = |x|$.

The absolute value function is one example of a piecewise-defined function. The "bracket" notation in Definition 1.9.7 is how we express which piece of the function applies on which interval. As we can see in Figure 1.9.8, for x values less than 0, the function $y = -x$ applies, whereas for x greater than or equal to 0, the rule is determined by $y = x$.

As long as we are careful to make sure that each potential input has one and only one corresponding output, we can define a piecewise function using as many different functions on different intervals as we desire.

Activity 1.9.4. In what follows, we work to understand two different piecewise functions entirely by hand based on familiar properties of linear and quadratic functions.

    a. Consider the function p defined by the following rule:

$$p(x) = \begin{cases} -(x+2)^2 + 2, & x < 0 \\ \frac{1}{2}(x-2)^2 + 1, & x \geq 0 \end{cases}$$

    What are the values of $p(-4)$, $p(-2)$, $p(0)$, $p(2)$, and $p(4)$?

    b. What point is the vertex of the quadratic part of p that is valid for $x < 0$? What point is the vertex of the quadratic part of p that is valid for $x \geq 0$?

    c. For what values of x is $p(x) = 0$? In addition, what is the y-intercept of p?

    d. Sketch an accurate, labeled graph of $y = p(x)$ on the axes provided in Figure 1.9.9.

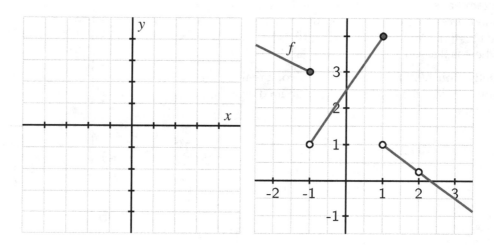

Figure 1.9.9: Axes to plot $y = p(x)$.   **Figure 1.9.10:** Graph of $y = f(x)$.

e. For the function f defined by Figure 1.9.10, determine a piecewise-defined formula for f that is expressed in bracket notation similar to the definition of $y = p(x)$ above.

1.9.4 Summary

- Just as we can generate a new number by adding, subtracting, multiplying, or dividing two given numbers, we can generate a new function by adding, subtracting, multiplying, or dividing two given functions. For instance, if we know formulas, graphs, or tables for functions f and g that share the same domain, we can create their product p according to the rule $p(x) = (f \cdot g)(x) = f(x) \cdot g(x)$.

- A piecewise function is a function whose formula consists of at least two different formulas in such a way that which formula applies depends on where the input falls in the domain. For example, given two functions f and g each defined on all real numbers, we can define a new piecewise function P according to the rule

$$P(x) = \begin{cases} f(x), & x < a \\ g(x), & x \geq a \end{cases}$$

This tells us that for any x to the left of a, we use the rule for f, whereas for any x to the right of or equal to a, we use the rule for g. We can use as many different functions as we want on different intervals, provided the intervals don't overlap.

1.9.5 Exercises

1.   For $f(x) = 2x + 4$ and $g(x) = 5x - 4$, find $(f - g)(x)$ and $(f - g)(-4)$.

2. For $f(t) = t - 2$ and $g(t) = t - 3$, find $\left(\dfrac{f}{g}\right)(t)$ and $\left(\dfrac{f}{g}\right)(-4)$.

3. For $f(x) = 2 + x^2$ and $g(x) = 2x - 1$, find $(f \cdot g)(x)$ and $(f \cdot g)(-3)$.

4.

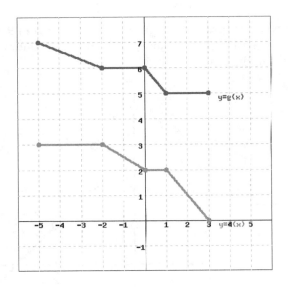
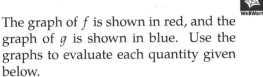

The graph of f is shown in red, and the graph of g is shown in blue. Use the graphs to evaluate each quantity given below.
a) $f(3)$
b) $g(3)$
c) $f(3) + g(3)$
d) $(f - g)(3)$

5.

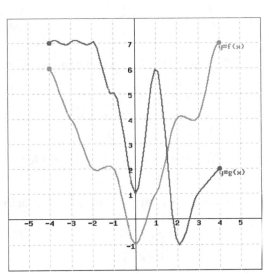

The graph of f is shown in red, and the graph of g is shown in blue. Use the graphs to evaluate each quantity given below.
a) $f(-2)$
b) $g(-2)$
c) $(f + g)(-2)$
d) $(g - f)(-2)$

6.

x	$f(x)$	$g(x)$
1	−3	2
2	3	4
3	1	−4
4	−4	−1
5	2	5

Use the table defining f and g to solve:
a) $(f - g)(4)$
b) $(f + g)(4) - (g - f)(5)$
c) $\left(\dfrac{f}{g}\right)(4)$

7. Let $r(t) = 2t - 3$ and $s(t) = 5 - 3t$. Determine a formula for each of the following new functions and simplify your result as much as possible.

    a. $f(t) = (r + s)(t)$

    d. $q(t) = (s \circ r)(t)$

    b. $g(t) = \left(\frac{s}{r}\right)(t)$

    c. $h(t) = (r \cdot s)(t)$

    e. $w(t) = r(t - 4) + 7$

8. Consider the functions s and g defined by the graphs in Figure 1.9.11 and Figure 1.9.12. Assume that to the left and right of the pictured domains, each function continues behaving according to the trends seen in the figures.

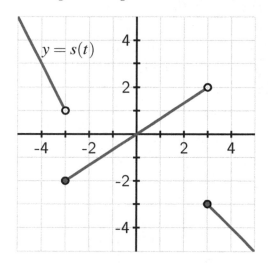

Figure 1.9.11: The graph of a piecewise function, s.

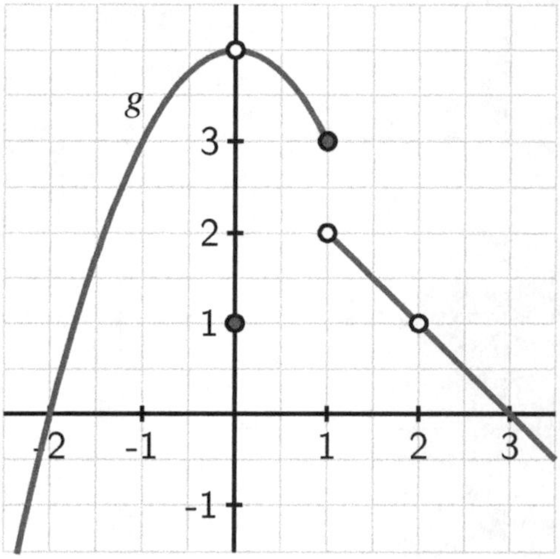

Figure 1.9.12: The graph of a piecewise function, g.

   a. Determine a piecewise formula for the function $y = s(t)$ that is valid for all real numbers t.

   b. Determine a piecewise formula for the function $y = g(x)$ that is valid for all real numbers x.

   c. Determine each of the following quantities or explain why they are not defined.

      i. $(s \cdot g)(1)$

      iii. $(s \circ g)(1.5)$

      ii. $(g - s)(3)$

      iv. $(g \circ s)(-4)$

9. One of the most important principles in the study of changing quantities is found in the relationship between distance, average velocity, and time. For a moving body traveling on a straight-line path at an average rate of v for a period of time t, the distance traveled, d, is given by

$$d = v \cdot t$$

    In the Ironman Triathlon, competitors swim 2.4 miles, bike 112 miles, and then run a 26.2 mile marathon. In the following sequence of questions, we build a piecewise function that models a competitor's location in the race at a given time t. To start, we have the following known information.

- She swims at an average rate of 2.5 miles per hour throughout the 2.4 miles in the water.

- Her transition from swim to bike takes 3 minutes (0.05 hours), during which time she doesn't travel any additional distance.

- She bikes at an average rate of 21 miles per hour throughout the 112 miles of biking.

- Her transition from bike to run takes just over 2 minutes (0.03 hours), during which time she doesn't travel any additional distance.

- She runs at an average rate of 8.5 miles per hour throughout the marathon.

- In the questions that follow, assume for the purposes of the model that the triathlete swims, bikes, and runs at essentially constant rates (given by the average rates stated above).

a. Determine the time the swimmer exits the water. Report your result in hours.

b. Likewise, determine the time the athlete gets off her bike, as well as the time she finishes the race.

c. List 5 key points in the form (time, distance): when exiting the water, when starting the bike, when finishing the bike, when starting the run, and when finishing the run.

d. What is the triathlete's average velocity over the course of the entire race? Is this velocity the average of her swim velocity, bike velocity, and run velocity? Why or why not?

e. Determine a piecewise function $s(t)$ whose value at any given time (in hours) is the triathlete's total distance traveled.

f. Sketch a carefully labeled graph of the triathlete's distance traveled as a function of time on the axes provided. Provide clear scale and note key points on the graph.

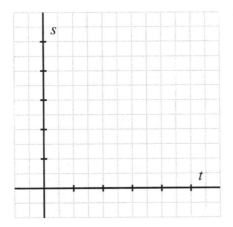 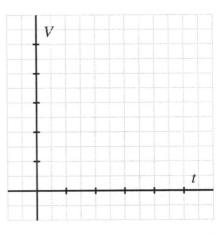

g. Sketch a possible graph of the triathete's *velocity*, V, as a function of time on the righthand axes. Here, too, label key points and provide clear scale. Write several sentences to explain and justify your graph.

CHAPTER 2

Circular Functions

2.1 Traversing Circles

Motivating Questions

- How does a point traversing a circle naturally generate a function?

- What are some important properties that characterize a function generated by a point traversing a circle?

- How does a circular function change in ways that are different from linear and quadratic functions?

Certain naturally occurring phenomena eventually repeat themselves, especially when the phenomenon is somehow connected to a circle. For example, suppose that you are taking a ride on a ferris wheel and we consider your height, h, above the ground and how your height changes in tandem with the distance, d, that you have traveled around the wheel. In Figure 2.1.1 we see a snapshot of this situation, which is available as a full animation[1] at http://gvsu.edu/s/0Dt.

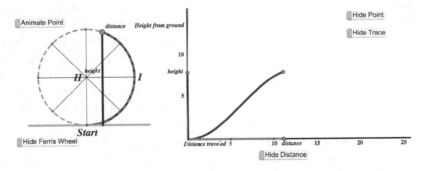

Figure 2.1.1: A snapshot of the motion of a cab moving around a ferris wheel. Reprinted with permission from Illuminations by the National Council of Teachers of Mathematics. All rights reserved.

[1]Used with permission from Illuminations by the National Council of Teachers of Mathematics. All rights reserved.

Because we have two quantities changing in tandem, it is natural to wonder if it is possible to represent one as a function of the other.

> **Preview Activity 2.1.1.** In the context of the ferris wheel pictured in Figure 2.1.1, assume that the height, h, of the moving point (the cab in which you are riding), and the distance, d, that the point has traveled around the circumference of the ferris wheel are both measured in meters.
>
> Further, assume that the circumference of the ferris wheel is 150 meters. In addition, suppose that after getting in your cab at the lowest point on the wheel, you traverse the full circle several times.
>
> a. Recall that the circumference, C, of a circle is connected to the circle's radius, r, by the formula $C = 2\pi r$. What is the radius of the ferris wheel? How high is the highest point on the ferris wheel?
>
> b. How high is the cab after it has traveled 1/4 of the circumference of the circle?
>
> c. How much distance along the circle has the cab traversed at the moment it first reaches a height of $\frac{150}{\pi} \approx 47.75$ meters?
>
> d. Can h be thought of as a function of d? Why or why not?
>
> e. Can d be thought of as a function of h? Why or why not?
>
> f. Why do you think the curve shown at right in Figure 2.1.1 has the shape that it does? Write several sentences to explain.

2.1.1 Circular Functions

The natural phenomenon of a point moving around a circle leads to interesting relationships. For easier arithmetic, let's consider a point traversing a circle of circumference 24 and examine how the point's height, h, changes as the distance traversed, d, changes. Note particularly that each time the point traverses $\frac{1}{8}$ of the circumference of the circle, it travels a distance of $24 \cdot \frac{1}{8} = 3$ units, as seen in Figure 2.1.2 where each noted point lies 3 additional units along the circle beyond the preceding one. Note that we know the exact heights of certain points. Since the circle has circumference $C = 24$, we know that $24 = 2\pi r$ and therefore $r = \frac{12}{\pi} \approx 3.82$. Hence, the point where $d = 6$ (located 1/4 of the way along the circle) is at a height of $h = \frac{12}{\pi} \approx 3.82$. Doubling this value, the point where $d = 12$ has height $h = \frac{24}{\pi} \approx 7.64$. Other heights, such as those that correspond to $d = 3$ and $d = 15$ (identified on the figure by the green line segments) are not obvious from the circle's radius, but can be estimated from the grid in Figure 2.1.2 as $h \approx 1.1$ (for $d = 3$) and $h \approx 6.5$ (for $d = 15$). Using all of these observations along with the symmetry of the circle, we can determine the other entries in Table 2.1.3. Moreover, if we now let the point continue traversing the circle, we observe that the d-values will increase accordingly, but the h-values will repeat according to the already-established pattern, resulting in the data in Table 2.1.4. It is apparent that each point on the circle corresponds to one and only one height, and thus we can view the

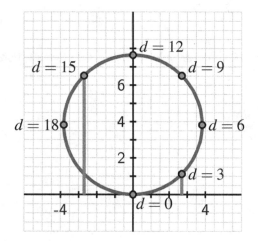

Figure 2.1.2: A point traversing a circle with circumference $C = 24$.

d	0	3	6	9	12	15	18	21	24
h	0	1.1	3.82	6.5	7.64	6.5	3.82	1.1	0

Table 2.1.3: Data for height, h, as a function of distance traversed, d.

height of a point as a function of the distance the point has traversed around the circle, say $h = f(d)$. Using the data from the two tables and connecting the points in an intuitive way, we get the graph shown in Figure 2.1.5. The function $h = f(d)$ we have been discussing is an example of what we will call a *circular function*. Indeed, it is apparent that if we

- take any circle in the plane,

- choose a starting location for a point on the circle,

- let the point traverse the circle continuously,

- and track the height of the point as it traverses the circle,

the height of the point is a function of distance traversed and the resulting graph will have the same basic shape as the curve shown in Figure 2.1.5. It also turns out that if we track the location of the x-coordinate of the point on the circle, the x-coordinate is also a function of distance traversed and its curve has a similar shape to the graph of the height of the point (the y-coordinate). Both of these functions are circular functions because they are generated by motion around a circle.

Activity 2.1.2. Consider the circle pictured in Figure 2.1.6 that is centered at the point $(2, 2)$ and that has circumference 8. Assume that we track the y-coordinate (that is, the height, h) of a point that is traversing the circle counterclockwise and that it starts at P_0 as pictured.

111

d	24	27	30	33	36	39	42	45	48
h	0	1.1	3.82	6.5	7.64	6.5	3.82	1.1	0

Table 2.1.4: Additional data for height, h, as a function of distance traversed, d.

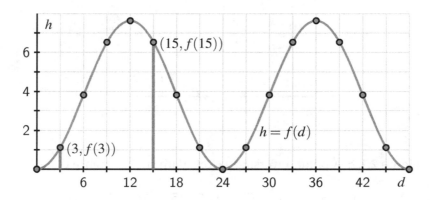

Figure 2.1.5: The height, h, of a point traversing a circle of radius 24 as a function of distance, d, traversed around the circle.

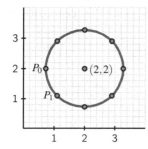

Figure 2.1.6: A point traversing the circle.

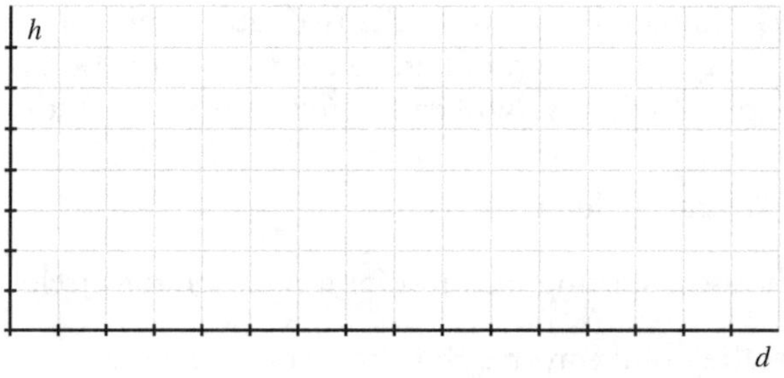

Figure 2.1.7: Axes for plotting h as a function of d.

a. How far along the circle is the point P_1 from P_0? Why?

b. Label the subsequent points in the figure P_2, P_3, ... as we move counterclockwise around the circle. What is the exact y-coordinate of the point P_2? of P_4? Why?

c. Determine the y-coordinates of the remaining points on the circle (exactly where possible, otherwise approximately) and hence complete the entries in Table 2.1.8 that track the height, h, of the point traversing the circle as a function of distance traveled, d. Note that the d-values in the table correspond to the point traversing the circle more than once.

d	0	1	2	3	4	5	6	7	8	9	10	11	12	13	14	15	16
h	2																

Table 2.1.8: Data for h as a function of d.

d. By plotting the points in Table 2.1.8 and connecting them in an intuitive way, sketch a graph of h as a function of d on the axes provided in Figure 2.1.7 over the interval $0 \leq d \leq 16$. Clearly label the scale of your axes and the coordinates of several important points on the curve.

e. What is similar about your graph in comparison to the one in Figure 2.1.5? What is different?

f. What will be the value of h when $d = 51$? How about when $d = 102$?

2.1.2 Properties of Circular Functions

Every circular function has several important features that are connected to the circle that defines the function. For the discussion that follows, we focus on circular functions that result from tracking the y-coordinate of a point traversing counterclockwise a circle of radius a centered at the point (k, m). Further, we will denote the circumference of the circle by the letter p.

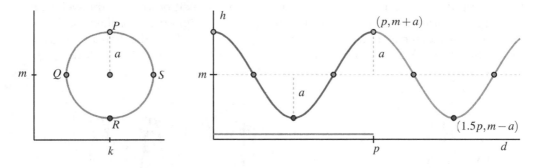

Figure 2.1.9: A point traversing the circle.

Figure 2.1.10: Plotting h as a function of d.

We assume that the point traversing the circle starts at P in Figure 2.1.9. Its height is initially $y = m + a$, and then its height decreases to $y = m$ as we traverse to Q. Continuing, the point's height falls to $y = m - a$ at R, and then rises back to $y = m$ at S, and eventually back up to $y = m + a$ at the top of the circle. If we plot these heights continuously as a function of distance, d, traversed around the circle, we get the curve shown at right in Figure 2.1.10. This curve has several important features for which we introduce important terminology.

The **midline** of a circular function is the horizontal line $y = m$ for which half the curve lies above the line and half the curve lies below. If the circular function results from tracking the y-coordinate of a point traversing a circle, $y = m$ corresponds to the y-coordinate of the center of the circle. In addition, the **amplitude** of a circular function is the maximum deviation of the curve from the midline. Note particularly that the value of the amplitude, a, corresponds to the radius of the circle that generates the curve.

Because we can traverse the circle in either direction and for as far as we wish, the domain of any circular function is the set of all real numbers. From our observations about the midline and amplitude, it follows that the range of a circular function with midline $y = m$ and amplitude a is the interval $[m - a, m + a]$.

Finally, we introduce the formal definition of a **periodic** function.

Definition 2.1.11 Let f be a function whose domain and codomain are each the set of all real numbers. We say that f is **periodic** provided that there exists a real number k such that $f(x + k) = f(x)$ for every possible choice of x. The smallest value p for which $f(x + p) = f(x)$ for every choice of x is called the **period** of f. ◊

For a circular function, the period is always the circumference of the circle that generates the curve. In Figure 2.1.10, we see how the curve has completed one full cycle of behavior every p units, regardless of where we start on the curve.

Circular functions arise as models for important phenomena in the world around us, such as in a *harmonic oscillator*. Consider a mass attached to a spring where the mass sits on a frictionless surface. After setting the mass in motion by stretching or compressing the spring, the mass will oscillate indefinitely back and forth, and its distance from a fixed point on the surface turns out to be given by a circular function.

Activity 2.1.3. A weight is placed on a frictionless table next to a wall and attached to a spring that is fixed to the wall. From its natural position of rest, the weight is imparted an initial velocity that sets it in motion. The weight then oscillates back and forth, and we can measure its distance, $h = f(t)$ (in inches) from the wall at any given time, t (in seconds). A graph of f and a table of select values are given below.

t	$f(t)$		t	$f(t)$
0.25	6.807		2.25	9.913
0.5	4.464		2.5	11.536
0.75	3.381		2.75	12.619
1	3.000		3	13.000
1.25	3.381		3.25	12.619
1.5	4.464		3.5	11.536
1.75	6.087		3.75	9.913
2	8.000		4	8.000

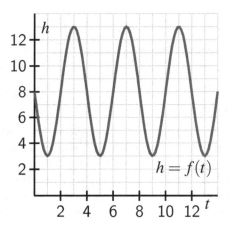

a. Determine the period p, midline $y = m$, and amplitude a of the function f.

b. What is the furthest distance the weight is displaced from the wall? What is the least distance the weight is displaced from the wall? What is the range of f?

c. Determine the average rate of change of f on the intervals $[4, 4.25]$ and $[4.75, 5]$. Write one careful sentence to explain the meaning of each (including units). In addition, write a sentence to compare the two different values you find and what they together say about the motion of the weight.

d. Based on the periodicity of the function, what is the value of $f(6.75)$? of $f(11.25)$?

2.1.3 The average rate of change of a circular function

Just as there are important trends in the values of a circular function, there are also interesting patterns in the average rate of change of the function. These patterns are closely tied to the geometry of the circle.

For the next part of our discussion, we consider a circle of radius 1 centered at $(0, 0)$, and consider a point that travels a distance d counterclockwise around the circle with its starting point viewed as $(1, 0)$. We use this circle to generate the circular function $h = f(d)$ that tracks the height of the point at the moment the point has traversed d units around the circle from $(1, 0)$. Let's consider the average rate of change of f on several intervals that are connected to certain fractions of the circumference.

Remembering that h is a function of distance traversed along the circle, it follows that the average rate of change of h on any interval of distance between two points P and Q on the circle is given by

$$AV_{[P,Q]} = \frac{\text{change in height}}{\text{distance along the circle}},$$

where both quantities are measured from point P to point Q.

First, in Figure 2.1.12, we consider points P, Q, and R where Q results from traversing $1/8$ of the circumference from P, and R $1/8$ of the circumference from Q. In particular, we note that the distance d_1 along the circle from P to Q is the same as the distance d_2 along the circle from Q to R, and thus $d_1 = d_2$. At the same time, it is apparent from the geometry of the circle that the change in height h_1 from P to Q is greater than the change in height h_2 from Q to R, so $h_1 > h_2$. Thus, we can say that

$$AV_{[P,Q]} = \frac{h_1}{d_1} > \frac{h_2}{d_2} = AV_{[Q,R]}.$$

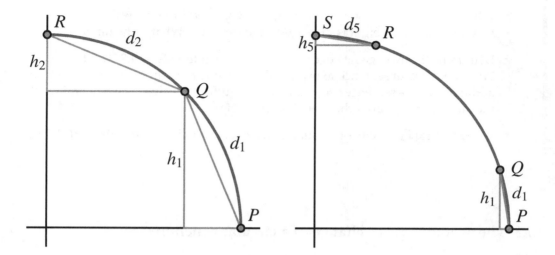

Figure 2.1.12: Comparing the average rate of change over 1/8 the circumference.

Figure 2.1.13: Comparing the average rate of change over 1/20 the circumference.

The differences in certain average rates of change appear to become more extreme if we consider shorter arcs along the circle. Next we consider traveling 1/20 of the circumference along the circle. In Figure 2.1.13, points P and Q lie 1/20 of the circumference apart, as do R and S, so here $d_1 = d_5$. In this situation, it is the case that $h_1 > h_5$ for the same reasons as above, but we can say even more. From the green triangle in Figure 2.1.13, we see that $h_1 \approx d_1$ (while $h_1 < d_1$), so that $AV_{[P,Q]} = \frac{h_1}{d_1} \approx 1$. At the same time, in the magenta triangle in the figure we see that h_5 is very small, especially in comparison to d_5, and thus $AV_{[R,S]} = \frac{h_5}{d_5} \approx 0$. Hence, in Figure 2.1.13,

$$AV_{[P,Q]} \approx 1 \text{ and } AV_{[R,S]} \approx 0.$$

This information tells us that a circular function appears to change most rapidly for points near its midline and to change least rapidly for points near its highest and lowest values.

We can study the average rate of change not only on the circle itself, but also on a graph such as Figure 2.1.10, and thus make conclusions about where the function is increasing, decreasing, concave up, and concave down.

Activity 2.1.4. Consider the same setting as Activity 2.1.3: a weight oscillates back and forth on a frictionless table with distance from the wall given by, $h = f(t)$ (in inches) at any given time, t (in seconds). A graph of f and a table of select values are given below.

t	f(t)	t	f(t)
0.25	6.807	2.25	9.913
0.5	4.464	2.5	11.536
0.75	3.381	2.75	12.619
1	3.000	3	13.000
1.25	3.381	3.25	12.619
1.5	4.464	3.5	11.536
1.75	6.087	3.75	9.913
2	8.000	4	8.000

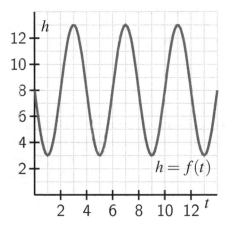

a. Determine $AV_{[2,2.25]}$, $AV_{[2.25,2.5]}$, $AV_{[2.5,2.75]}$, and $AV_{[2.75,3]}$. What do these four values tell us about how the weight is moving on the interval $[2, 3]$?

b. Give an example of an interval of length 0.25 units on which f has its most negative average rate of change. Justify your choice.

c. Give an example of the longest interval you can find on which f is decreasing.

d. Give an example of an interval on which f is concave up.[2]

e. On an interval where f is both decreasing and concave down, what does this tell us about how the weight is moving on that interval? For instance, is the weight moving toward or away from the wall? is it speeding up or slowing down?

f. What general conclusions can you make about the average rate of change of a circular function on intervals near its highest or lowest points? about its average rate of change on intervals near the function's midline?

2.1.4 Summary

• When a point traverses a circle, a corresponding function can be generated by tracking the height of the point as it moves around the circle, where height is viewed as a function of distance traveled around the circle. We call such a function a *circular function*. An image that shows how a circular function's graph is generated from the circle can be seen in Figure 2.1.10.

• Circular functions have several standard features. The function has a *midline* that is the line for which half the points on the curve lie above the line and half the points on the curve lie below. A circular function's *amplitude* is the maximum deviation of the

---

[2]Recall that a function is concave up on an interval provided that throughout the interval, the curve bends upward, similar to a parabola that opens up.

function value from the midline; the amplitude corresponds to the radius of the circle that generates the function. Circular functions also repeat themselves, and we call the smallest value of p for which $f(x + p) = f(x)$ for all x the period of the function. The period of a circular function corresponds to the circumference of the circle that generates the function.

- Non-constant linear functions are either always increasing or always decreasing; quadratic functions are either always concave up or always concave down. Circular functions are sometimes increasing and sometimes decreasing, plus sometimes concave up and sometimes concave down. These behaviors are closely tied to the geometry of the circle.

2.1.5 Exercises

1. Let $y = f(x)$ be a periodic function whose values are given below. Find the period, amplitude, and midline.

x	5	25	45	65	85	105	125	145	165
f(x)	17	15	-3	17	15	-3	17	15	-3

2. A ferris wheel is 140 meters in diameter and boarded at its lowest point (6 O'Clock) from a platform which is 8 meters above ground. The wheel makes one full rotation every 14 minutes, and at time $t = 0$ you are at the loading platform (6 O'Clock). Let $h = f(t)$ denote your height above ground in meters after t minutes.

    (a) What is the period of the function $h = f(t)$?

    (b) What is the midline of the function $h = f(t)$?

    (c) What is the amplitude of the function $h = f(t)$?

    (d) Consider the six possible graphs of $h = f(t)$ below. Be sure to carefully read the labels on the axes in order distinguish the key features of each graph.

    Which (if any) of the graphs A-F represents two full revolutions of the ferris wheel described above?

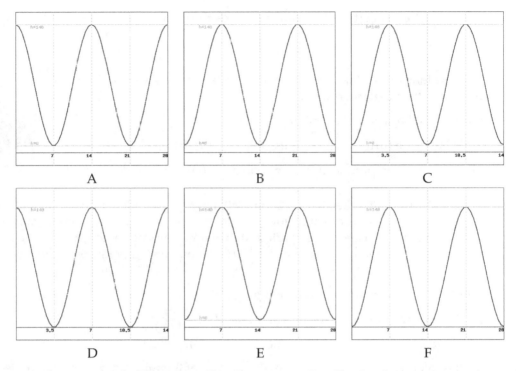

<table>
<tr><td>A</td><td>B</td><td>C</td></tr>
<tr><td>D</td><td>E</td><td>F</td></tr>
</table>

3. A weight is suspended from the ceiling by a spring. Let d be the distance in centimeters from the ceiling to the weight. When the weight is motionless, $d = 11$ cm. If the weight is disturbed, it begins to bob up and down, or *oscillate*. Then d is a periodic function of t, the time in seconds, so $d = f(t)$. Consider the graph of $d = f(t)$ below, which represents the distance of the weight from the ceiling at time t.

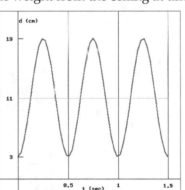

(a) Based on the graph of $d = f(t)$ above, which of the statements below correctly describes the motion of the weight as it bobs up and down?

⊙ The weight starts closest to the floor and begins by bouncing up towards the ceiling.

⊙ The weight starts closest to the ceiling and begins by stretching the spring down towards the floor.

⊙ The spring starts at its average distance between the ceiling and floor and begins by stretching the spring down towards the floor.

⊙ None of the above

(b) How long does it take the weight to bounce completely up and down (or down and up) and return to its starting position?

(c) What is the closest the weight gets to the ceiling?

(d) What is the furthest the weights gets from the ceiling?

(e) What is the amplitdue of the graph of $d = f(t)$?

4. The temperature of a chemical reaction oscillates between a low of $10\,°C$ and a high of $135\,°C$. The temperature is at its lowest point at time $t = 0$, and reaches its maximum point over a two and a half hour period. It then takes the same amount of time to return back to its initial temperature. Let $y = H(t)$ denote the temperature of the reaction t hours after the reaction begins.

(a) What is the period of the function $y = H(t)$?

(b) What is the midline of the function $y = H(t)$?

(c) What is the amplitude of the function $y = H(t)$?

(d) Based on your answers above, make a graph of the function $y = H(t)$ on a piece of paper. Which of the graphs below best matches your graph?

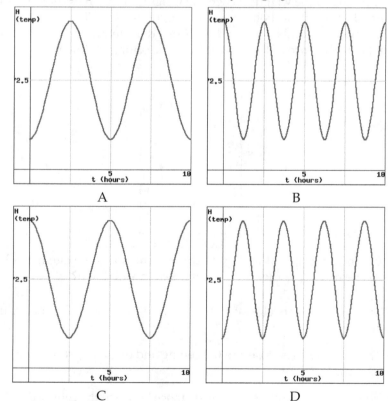

5. Consider the circle pictured in Figure 2.1.14 that is centered at the point $(2,2)$ and that has circumference 8. Suppose that we track the x-coordinate (that is, the horizontal location, which we will call k) of a point that is traversing the circle counterclockwise and that it starts at P_0 as pictured.

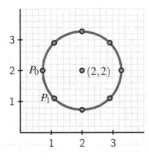
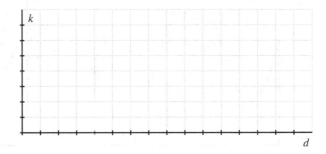

Figure 2.1.14: A point traversing the circle.

Figure 2.1.15: Axes for plotting k as a function of d.

Recall that in Activity 2.1.2 we identified the exact and approximate vertical coordinates of all 8 noted points on the unit circle. In addition, recall that the radius of the circle is $r = \frac{8}{2\pi} \approx 1.2732$.

a. What is the exact horizontal coordinate of P_0? Why?

b. Complete the entries in Table 2.1.16 that track the horizontal location, k, of the point traversing the circle as a function of distance traveled, d.

d	0	1	2	3	4	5	6	7	8	9	10	11	12	13	14	15	16
k	0.73																

Table 2.1.16: Data for h as a function of d.

c. By plotting the points in Table 2.1.16 and connecting them in an intuitive way, sketch a graph of k as a function of d on the axes provided in Figure 2.1.15 over the interval $0 \le d \le 16$. Clearly label the scale of your axes and the coordinates of several important points on the curve.

d. What is similar about your graph in comparison to the one in Figure 2.1.7? What is different?

e. What will be the value of k when $d = 51$? How about when $d = 102$?

6. Two circular functions, f and g, are generated by tracking the y-coordinate of a point traversing two different circles. The resulting graphs are shown in Figure 2.1.17 and Figure 2.1.18. Assuming the horizontal scale matches the vertical scale, answer the following questions for each of the functions f and g.

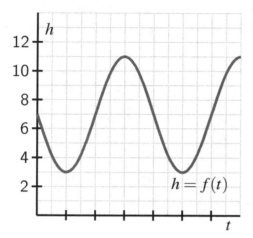

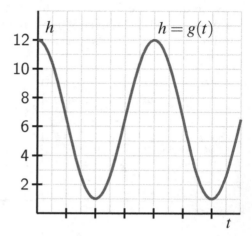

Figure 2.1.17: A plot of the circular function f.

Figure 2.1.18: A plot of the circular function g.

    a. Assume that the circle used to generate the circular function is centered at the point $(0, m)$ and has radius r. What are the numerical values of m and r? Why?

    b. What are the coordinates of the location on the circle at which the point begins its traverse? Said differently, what point on the circle corresponds to $t = 0$ on the function's graph?

    c. What is the period of the function? How is this connected to the circle and to the scale on the horizontal axes on which the function is graphed?

    d. How would the graph look if the circle's radius was 1 unit larger? 1 unit smaller?

7.   A person goes for a ride on a ferris wheel. They enter one of the cars at the lowest possible point on the wheel from a platform 7 feet off the ground. When they are at the very top of the wheel, they are 92 feet off the ground. Let h represent the height of the car (in feet) and d (in feet) the distance the car has traveled along the wheel's circumference from its starting location at the bottom of the wheel. We'll use the notation $h = f(d)$ for how height is a function of distance traveled.

    a. How high above the ground is the center of the ferris wheel?

    b. How far does the car travel in one complete trip around the wheel?

    c. For the circular function $h = f(d)$, what is its amplitude? midline? period?

    d. Sketch an accurate graph of h through at least two full periods. Clearly label the scale on the horizontal and vertical axes along with several important points.

2.2 The Unit Circle

Motivating Questions

- What is the radian measure of an angle?

- Are there natural special points on the unit circle whose coordinates we can identify exactly?

- How can we determine arc length and the location of special points in circles other than the unit circle?

As demonstrated by several different examples in Section 2.1, certain periodic phenomena are closely linked to circles and circular motion. Rather than regularly work with circles of different center and radius, it turns out to be ideal to work with one standard circle and build all circular functions from it. The **unit circle** is the circle of radius 1 that is centered at the origin, $(0, 0)$.

If we pick any point (x, y) that lies on the unit circle, the point is associated with a right triangle whose horizontal leg has length $|x|$ and whose vertical leg has length $|y|$, as seen in Figure 2.2.1. By the Pythagorean Theorem, it follows that

$$x^2 + y^2 = 1,$$

and this is the equation of the unit circle: a point (x, y) lies on the unit circle if and only if $x^2 + y^2 = 1$.

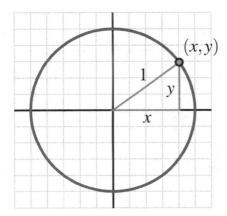

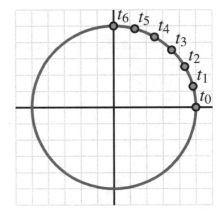

Figure 2.2.1: Coordinates of a point on the unit circle.

Figure 2.2.2: A point traversing the unit circle.

To study the circular functions generated by the unit circle, we will also animate a point and let it traverse the circle. Starting at $(1, 0)$ indicated by t_0 in Figure 2.2.2, we see a sequence of points that result from traveling a distance along the circle that is $1/24$ the circumference of the unit circle. Since the unit circle's circumference is $C = 2\pi r = 2\pi$, it follows that the distance from t_0 to t_1 is

$$d = \frac{1}{24} \cdot 2\pi = \frac{\pi}{12}.$$

As we work to better understand the unit circle, we will commonly use fractional multiples of π as these result in natural distances traveled along the unit circle.

Preview Activity 2.2.1. In Figure 2.2.3 there are 24 equally spaced points on the unit circle. Since the circumference of the unit circle is 2π, each of the points is $\frac{1}{24} \cdot 2\pi = \frac{\pi}{12}$ units apart (traveled along the circle). Thus, the first point counterclockwise from $(1,0)$ corresponds to the distance $t = \frac{\pi}{12}$ traveled along the unit circle. The second point is twice as far, and thus $t = 2 \cdot \frac{\pi}{12} = \frac{\pi}{6}$ units along the circle away from $(1,0)$.

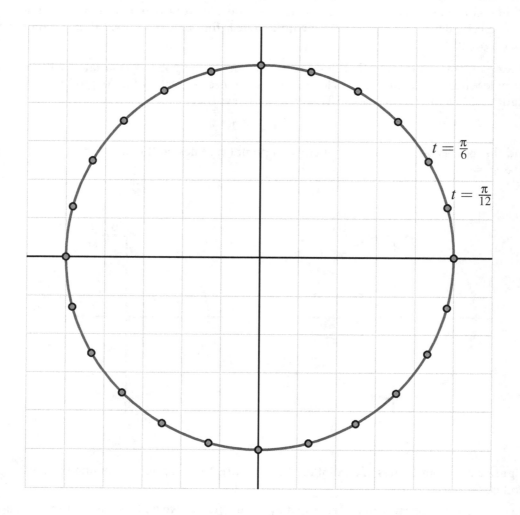

Figure 2.2.3: The unit circle with 24 equally-spaced points.

a. Label each of the subsequent points on the unit circle with the exact distance they lie counter-clockwise away from $(1, 0)$; write each fraction in lowest terms.

b. Which distance along the unit circle corresponds to $\frac{1}{4}$ of a full rotation around? to $\frac{5}{8}$ of a full rotation?

c. One way to measure angles is connected to the arc length along a circle. For an angle whose vertex is at $(0, 0)$ in the unit circle, we say the angle's measure is 1 **radian** provided that the angle intercepts an arc of the circle that is 1 unit in length, as pictured in Figure 2.2.4. Note particularly that an angle measuring 1 radian intercepts an arc of the same length as the circle's radius.

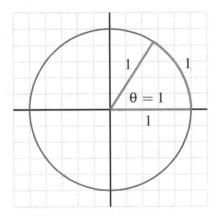

Figure 2.2.4: An angle θ of measure 1 radian.

Suppose that α and β are each central angles and that their respective radian measures are $\alpha = \frac{\pi}{3}$ and $\beta = \frac{3\pi}{4}$. Sketch the angles α and β on the unit circle in Figure 2.2.3.

d. What is the radian measure that corresonds to a $90°$ angle?

2.2.1 Radians and degrees

In Preview Activity 2.2.1, we introduced the idea of radian measure of an angle. Here we state the formal definition of this term.

Definition 2.2.5 An angle whose vertex is at the center of a circle[1] measures 1 **radian** provided that the arc the angle intercepts on the circle equals the radius of the circle. ◊

As seen in Figure 2.2.4, in the unit circle this means that a central angle has measure 1 radian whenever it intercepts an arc of length 1 unit along the circumference. Because of this important correspondence between the unit circle and radian measure (one unit of arc length on the unit circle corresponds to 1 radian), we focus our discussion of radian measure within the unit circle.

Since there are 2π units of length along the unit circle's circumference it follows there are $\frac{1}{4} \cdot 2\pi = \frac{\pi}{2}$ units of length in $\frac{1}{4}$ of a revolution. We also know that $\frac{1}{4}$ of a revolution corresponds to a central angle that is a right angle, whose familiar degree measure is $90°$. If we extend

---

[1]We often call such an angle a *central angle*.

to a central angle that intercepts half the circle, we see similarly that π radians corresponds to 180°; this relationship enables us to convert angle measures from radians to degrees and vice versa.

Converting between radians and degrees.

An angle whose radian measure is 1 radian has degree measure $\frac{180}{\pi}°$. An angle whose degree measure is 1° has radian measure $\frac{\pi}{180}$.

Activity 2.2.2. Convert each of the following quantities to the alternative measure: degrees to radians or radians to degrees.

a. 30°

b. $\frac{2\pi}{3}$ radians

c. $\frac{5\pi}{4}$ radians

d. 240°

e. 17°

f. 2 radians

Note that in Figure 2.2.3 in the Preview Activity, we labeled 24 equally spaced points with their respective distances around the unit circle counterclockwise from $(1, 0)$. Because these distances are on the unit circle, they also correspond to the radian measure of the central angles that intercept them. In particular, each central angle with one of its sides on the positive x-axis generates a unique point on the unit circle, and with it, an associated length intercepted along the circumference of the circle. A good exercise at this point is to return to Figure 2.2.3 and label each of the noted points with the degree measure that is intercepted by a central angle with one side on the positive x-axis, in addition to the arc lengths (radian measures) already identified.

2.2.2 Special points on the unit circle

Our in-depth study of the unit circle is motivated by our desire to better understand the behavior of circular functions. Recall that as we traverse a circle, the height of the point moving along the circle generates a function that depends on distance traveled along the circle. Wherever possible, we'd like to be able to identify the exact height of a given point on the unit circle. Two special right triangles enable us to locate exactly an important collection of points on the unit circle.

Activity 2.2.3. In what follows, we work to understand key relationships in 45°-45°-90° and 30°-60°-90° triangles.

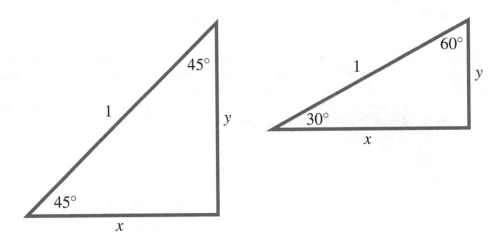

Figure 2.2.6: A right triangle with two 45° angles.

Figure 2.2.7: A right triangle with a 30° angle.

a. For the 45°-45°-90° triangle with legs x and y and hypotenuse 1, what does the fact that the triangle is isosceles tell us about the relationship between x and y? What are their exact values?

b. Now consider the 30°-60°-90° triangle with hypotenuse 1 and the longer leg lying along the positive x-axis. What special kind of triangle is formed when we reflect this triangle across the x-axis? How can we use this perspective to determine the exact values of x and y?

c. Suppose we consider the related 30°-60°-90° triangle with hypotenuse 1 and the shorter leg lying along the positive x-axis. What are the *exact* values of x and y in this triangle?

d. We know from the conversion factor from degrees to radians that an angle of 30° corresponds to an angle measuring $\frac{\pi}{6}$ radians, an angle of 45° corresponds to $\frac{\pi}{4}$ radians, and 60° corresponds to $\frac{\pi}{3}$ radians.

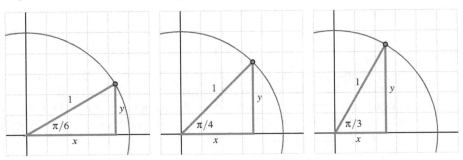

Figure 2.2.8: An angle measuring $\frac{\pi}{6}$ radians.

Figure 2.2.9: An angle measuring $\frac{\pi}{6}$ radians.

Figure 2.2.10: An angle measuring $\frac{\pi}{6}$ radians.

Use your work in (a), (b), and (c) to label the noted point in each of Figure 2.2.8, Figure 2.2.9, and Figure 2.2.10, respectively, with its exact coordinates.

Our work in Activity 2.2.3 enables us to identify exactly the location of 12 special points on the unit circle. In part (d) of the activity, we located the three noted points in Figure 2.2.11 along with their respective radian measures. By symmetry across the coordinate axes and thinking about the signs of coordinates in the other three quadrants, we can now identify all of the coordinates of the remaining 9 points.

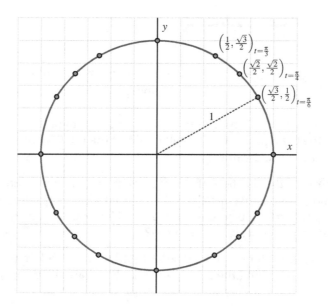

Figure 2.2.11: The unit circle with 16 special points whose location we can determine exactly.

In addition, we note that there are four additional points on the circle that we can locate exactly: the four points that correspond to angle measures of 0, $\frac{\pi}{2}$, π, and $\frac{3\pi}{2}$ radians, which lie where the coordinate axes intersect the circle. Each such point has 0 for one coordinate and ± 1 for the other. Labeling all of the remaining points in Figure 2.2.11 is an important exercise that you should do on your own.

Finally, we note that we can identify any point on the unit circle exactly simply by choosing one of its coordinates. Since every point (x, y) on the unit circle satisfies the equation $x^2 + y^2 = 1$, if we know the value of x or y and the quadrant in which the point lies, we can determine the other coordinate exactly.

2.2.3 Special points and arc length in non-unit circles

All of our work with the unit circle can be extended to circles centered at the origin with different radii, since a circle with a larger or smaller radius is a scaled version of the unit circle. For instance, if we instead consider a circle of radius 7, the coordinates of every point on the unit circle are magnified by a factor of 7, so the point that corresponds to an angle such as $\theta = \frac{2\pi}{3}$ has coordinates $\left(-\frac{7}{2}, \frac{7\sqrt{3}}{2}\right)$. Distance along the circle is magnified by the same factor: the arc length along the unit circle from $(0,0)$ to $\left(-\frac{7}{2}, \frac{7\sqrt{3}}{2}\right)$ is $7 \cdot \frac{2\pi}{3}$, since the arc length along the unit circle for this angle is $\frac{2\pi}{3}$.

If we think more generally about a circle of radius r with a central angle θ that intercepts an arc of length s, we see how the magnification factor r (in comparison to the unit circle) connects arc length and the central angle according to the following principle.

Connecting arc length and angles in non-unit circles.

If a central angle measuring θ radians intercepts an arc of length s in a circle of radius r, then

$$s = r\theta.$$

In the unit circle, where $r = 1$, the equation $s = r\theta$ demonstrates the familiar fact that arc length matches the radian measure of the central angle. Moreover, we also see how this formula aligns with the definition of radian measure: if the arc length and radius are equal, then the angle measures 1 radian.

Activity 2.2.4. Determine each of the following values or points exactly.

   a. In a circle of radius 11, the arc length intercepted by a central angle of $\frac{5\pi}{3}$.

   b. In a circle of radius 3, the central angle measure that intercepts an arc of length $\frac{\pi}{4}$.

   c. The radius of the circle in which an angle of $\frac{7\pi}{6}$ intercepts an arc of length $\frac{\pi}{2}$.

   d. The exact coordinates of the point on the circle of radius 5 that lies $\frac{25\pi}{6}$ units counterclockwise along the circle from $(5,0)$.

2.2.4 Summary

- The radian measure of an angle connects the measure of a central angle in a circle to the radius of the circle. A central angle has radian measure 1 provided that it intercepts an arc of length equal to the circle's radius. In the unit circle, a central angle's radian measure is precisely the same numerical value as the length of the arc it intercepts along the circle.

- If we begin at the point $(1,0)$ and move counterclockwise along the unit circle, there are natural special points on the unit circle that correspond to angles of measure $30°$, $45°$,

60°, and their multiples. We can count in 30° increments and identify special points that correspond to angles of measure 30°, 60°, 90°, 120°, and so on; doing likewise with 45°, these correspond to angles of 45°, 90°, 135°, etc. In radian measure, these sequences together give us the important angles $\frac{\pi}{6}, \frac{\pi}{4}, \frac{\pi}{3}, \frac{\pi}{2}, \frac{2\pi}{3}, \frac{3\pi}{4}$, and so on. Together with our work involving 45°-45°-90° and 30°-60°-90° triangles in Activity 2.2.3, we are able to identify the exact locations of all of the points in Figure 2.2.11.

- In any circle of radius r, if a central angle of measure θ radians intercepts an arc of length s, then it follows that
$$s = r\theta.$$
This shows that arc length, s, is magnified along with the size of the radius, r, of the circle.

2.2.5 Exercises

 1. Find t for the following terminal points:

(a) $P(0, 1)$

(b) $P(-\frac{\sqrt{2}}{2}, \frac{\sqrt{2}}{2})$

(c) $P(-1, 0)$

(d) $P(-\frac{\sqrt{2}}{2}, -\frac{\sqrt{2}}{2})$

(e) $P(0, -1)$

(f) $P(\frac{\sqrt{2}}{2}, -\frac{\sqrt{2}}{2})$

 2. What is the exact radian angle measure for 45° as a fraction of π?

What is a decimal approximation for the radian angle measure for 45° accurate to three decimal places?

 3. What is the exact radian angle measure for 22° as a fraction of π?

What is a decimal approximation for the radian angle measure for 22° accurate to three decimal places?

 4. Find the length of an arc on a circle of radius 6 corresponding to an angle of 90°.

 5. What is the length of an arc cut off by an angle of 2.5 radians on a circle of radius 4 inches?

 6. What angle (in degrees) corresponds to 17.7 rotations around the unit circle?

 7. An angle of $\dfrac{5\pi}{6}$ radians can be converted to an angle of _____ degrees.

8. Let (x, y) by a point on the unit circle. In each of the following situations, determine requested value exactly.

a. Suppose that $x = -0.3$ and y is negative. Find the value of y.

b. Suppose that (x, y) lies in Quadrant II and $x = -2y$. Find the values of x and y.

c. Suppose that (x, y) lies a distance of $\frac{29\pi}{6}$ units clockwise around the circle from

$(1, 0)$. Find the values of x and y.

    d. At what exact point(s) does the line $y = \frac{1}{2}x + \frac{1}{2}$ intersect the unit circle?

9. The unit circle is centered at $(0,0)$ and radius $r = 1$, from which the Pythagorean Theorem tells us that any point (x, y) on the unit circle satisfies the equation $x^2 + y^2 = 1$.

    a. Explain why any point (x, y) on a circle of radius r centered at (h, k) satisfies the equation $(x - h)^2 + (y - k)^2 = r^2$.

    b. Determine the equation of a circle centered at $(-3, 5)$ with radius $r = 2$.

    c. Suppose that the unit circle is magnified by a factor of 5 and then shifted 4 units right and 7 units down. What is the equation of the resulting circle?

    d. What is the length of the arc intercepted by a central angle of $\frac{2\pi}{3}$ radians in the circle $(x - 1)^2 + (y - 3)^2 = 16$?

    e. Suppose that the line segment from $(-2, -1)$ to $(4, 2)$ is a diameter of a circle. What is the circle's center, radius, and equation?

10. Consider the circle whose center is $(0, 0)$ and whose radius is $r = 5$. Let a point (x, y) traverse the circle counterclockwise from $(5, 0)$, and say the distance along the circle from $(5, 0)$ is represented by d.

    a. Consider the point (a, b) that is generated by the central angle θ with vertices $(5, 0)$, $(0, 0)$, and (a, b). If $\theta = \frac{\pi}{6}$, what are the exact values of a and b?

    b. Answer the same question as in (a) except with $\theta = \frac{\pi}{4}$ and $\theta = \frac{\pi}{3}$.

    c. How far has the point (x, y) traveled after it has traversed the circle one full revolution?

    d. Let $h = f(d)$ be the circular function that tracks the height of the point (x, y) as a function of distance, d, traversed counterclockwise from $(5, 0)$. Sketch an accurate graph of f through two full periods, labeling several special points on the graph as well as the horizontal and vertical scale of the axes.

2.3 The Sine and Cosine Functions

Motivating Questions

- What are the sine and cosine functions and how do they arise from a point traversing the unit circle?

- What important properties do the sine and cosine functions share?

- How do we compute values of $\sin(t)$ and $\cos(t)$, either exactly or approximately?

In Section 2.1, we saw how tracking the height of a point that is traversing a cirle generates a periodic function, such as in Figure 2.1.10. Then, in Section 2.2, we identified a collection of 16 special points on the unit circle, as seen in Figure 2.3.1.

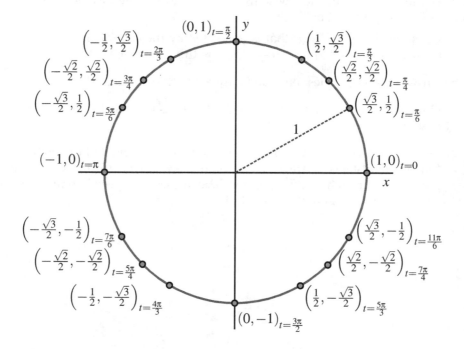

Figure 2.3.1: The unit circle with 16 labeled special points.

You can also use the *Desmos* file at http://gvsu.edu/s/0xt to review and study the special points on the unit circle.

Preview Activity 2.3.1. If we consider the unit circle in Figure 2.3.1, start at $t = 0$, and traverse the circle counterclockwise, we may view the height, h, of the traversing point as a function of the angle, t, in radians. From there, we can plot the resulting (t, h) ordered pairs and connect them to generate the circular function pictured in Figure 2.3.2.

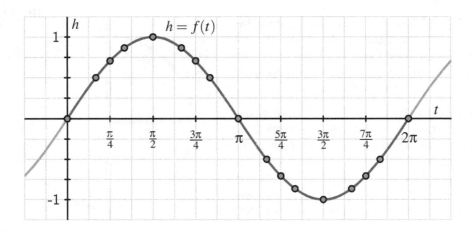

Figure 2.3.2: Plot of the circular function that tracks the height of a point traversing the unit circle.

a. What is the exact value of $f(\frac{\pi}{4})$? of $f(\frac{\pi}{3})$?

b. Complete the following table with the exact values of h that correspond to the stated inputs.

t	0	$\frac{\pi}{6}$	$\frac{\pi}{4}$	$\frac{\pi}{3}$	$\frac{\pi}{2}$	$\frac{2\pi}{3}$	$\frac{3\pi}{4}$	$\frac{5\pi}{6}$	π
h									

t	π	$\frac{7\pi}{6}$	$\frac{5\pi}{4}$	$\frac{4\pi}{3}$	$\frac{3\pi}{2}$	$\frac{5\pi}{3}$	$\frac{7\pi}{4}$	$\frac{11\pi}{6}$	2π
h									

Table 2.3.3: Exact values of h as a function of t.

c. What is the exact value of $f(\frac{11\pi}{4})$? of $f(\frac{14\pi}{3})$?

d. Give four different values of t for which $f(t) = -\frac{\sqrt{3}}{2}$.

2.3.1 The definition of the sine function

The circular function that tracks the height of a point on the unit circle traversing counter-clockwise from $(1,0)$ as a function of the corresponding central angle (in radians) is one of the most important functions in mathematics. As such, we give the function a name: the **sine** function.

Definition 2.3.4 Given a central angle in the unit circle that measures t radians and that intersects the circle at both $(1,0)$ and (a,b), as shown in Figure 2.3.5, we define the **sine of** t, denoted $\sin(t)$, by the rule $\sin(t) = b$. ◊

Because of the correspondence between radian angle measure and distance traversed on the unit circle, we can also think of $\sin(t)$ as identifying the y-coordinate of the point after it has traveled t units counterclockwise along the circle from $(1,0)$. Note particularly that we can consider the sine of negative inputs: for instance, $\sin(-\frac{\pi}{2}) = -1$.

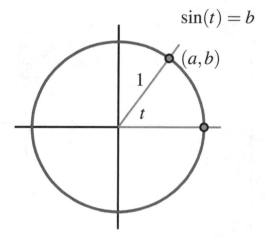

Figure 2.3.5: The definition of the sine of an angle t.

Based on our earlier work with the unit circle, we know many different exact values of the sine function, and summarize these in Table 2.3.6.

t	0	$\frac{\pi}{6}$	$\frac{\pi}{4}$	$\frac{\pi}{3}$	$\frac{\pi}{2}$	$\frac{2\pi}{3}$	$\frac{3\pi}{4}$	$\frac{5\pi}{6}$	π
$\sin(t)$	0	$\frac{1}{2}$	$\frac{\sqrt{2}}{2}$	$\frac{\sqrt{3}}{2}$	1	$\frac{\sqrt{3}}{2}$	$\frac{\sqrt{2}}{2}$	$\frac{1}{2}$	0

t	π	$\frac{7\pi}{6}$	$\frac{5\pi}{4}$	$\frac{4\pi}{3}$	$\frac{3\pi}{2}$	$\frac{5\pi}{3}$	$\frac{7\pi}{4}$	$\frac{11\pi}{6}$	2π
$\sin(t)$	0	$-\frac{1}{2}$	$-\frac{\sqrt{2}}{2}$	$-\frac{\sqrt{3}}{2}$	-1	$-\frac{\sqrt{3}}{2}$	$-\frac{\sqrt{2}}{2}$	$-\frac{1}{2}$	0

Table 2.3.6: Values of $h(t) = \sin(t)$ at special points on the unit circle.

Moreover, if we now plot these points in the usual way, as we did in Preview Activity 2.3.1, we get the familiar circular wave function that comes from tracking the height of a point traversing a circle. We often call the the graph in Figure 2.3.7 the *sine wave*.

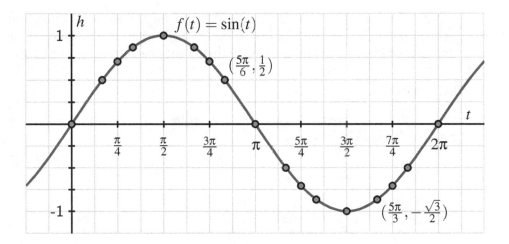

Figure 2.3.7: Plot of the sine function on the interval $[-\frac{\pi}{4}, \frac{7\pi}{4}]$.

2.3.2 The definition of the cosine function

Given any central angle of radian measure t in the unit circle with one side passing through the point $(1, 0)$, the angle generates a unique point (a, b) that lies on the circle. Just as we can view the y-coordinate as a function of t, the x-coordinate is likewise a function of t. We therefore make the following definition.

Definition 2.3.8 Given a central angle in the unit circle that measures t radians and that intersects the circle at both $(1, 0)$ and (a, b), as shown in Figure 2.3.9, we define the **cosine of** t, denoted $\cos(t)$, by the rule $\cos(t) = a$. ◊

Again because of the correspondence between the radian measure of an angle and arc length along the unit circle, we can view the value of $\cos(t)$ as tracking the x-coordinate of a point traversing the unit circle clockwise a distance of t units along the circle from $(1, 0)$. We now use the data and information we have developed about the unit circle to build a table of values of $\cos(t)$ as well as a graph of the curve it generates.

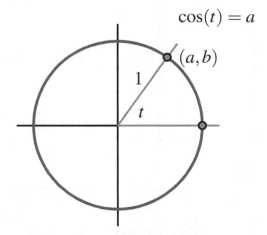

Figure 2.3.9: The definition of the cosine of an angle t.

135

Activity 2.3.2. Let $k = g(t)$ be the function that tracks the x-coordinate of a point traversing the unit circle counterclockwise from $(1,0)$. That is, $g(t) = \cos(t)$. Use the information we know about the unit circle that is summarized in Figure 2.3.1 to respond to the following questions.

a. What is the exact value of $\cos(\frac{\pi}{6})$? of $\cos(\frac{5\pi}{6})$? $\cos(-\frac{\pi}{3})$?

b. Complete the following table with the exact values of k that correspond to the stated inputs.

t	0	$\frac{\pi}{6}$	$\frac{\pi}{4}$	$\frac{\pi}{3}$	$\frac{\pi}{2}$	$\frac{2\pi}{3}$	$\frac{3\pi}{4}$	$\frac{5\pi}{6}$	π
k									

t	π	$\frac{7\pi}{6}$	$\frac{5\pi}{4}$	$\frac{4\pi}{3}$	$\frac{3\pi}{2}$	$\frac{5\pi}{3}$	$\frac{7\pi}{4}$	$\frac{11\pi}{6}$	2π
k									

Table 2.3.10: Exact values of $k = g(t) = \cos(t)$.

c. On the axes provided in Figure 2.3.11, sketch an accurate graph of $k = \cos(t)$. Label the exact location of several key points on the curve.

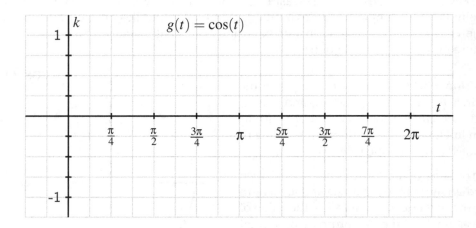

Figure 2.3.11: Axes for plotting $k = \cos(t)$.

d. What is the exact value of $\cos(\frac{11\pi}{4})$? of $\cos(\frac{14\pi}{3})$?

e. Give four different values of t for which $\cos(t) = -\frac{\sqrt{3}}{2}$.

f. How is the graph of $k = \cos(t)$ different from the graph of $h = \sin(t)$? How are the graphs similar?

As we work with the sine and cosine functions, it's always helpful to remember their definitions in terms of the unit circle and the motion of a point traversing the circle. At http://gvsu.edu/s/0xe you can explore and investigate a helpful *Desmos* animation that shows how this motion around the circle generates each of the respective graphs.

2.3.3 Properties of the sine and cosine functions

Because the sine function results from tracking the y-coordinate of a point traversing the unit circle and the cosine function from the x-coordinate, the two functions have several shared properties of circular functions.

Properties of the sine and cosine functions.

For both $f(t) = \sin(t)$ and $g(t) = \cos(t)$,

- the domain of the function is all real numbers;

- the range of the function is $[-1, 1]$;

- the midline of the function is $y = 0$;

- the amplitude of the function is $a = 1$;

- the period of the function is $p = 2\pi$.

It is also insightful to juxtapose the sine and cosine functions' graphs on the same coordinate axes. When we do, as seen in Figure 2.3.12, we see that the curves can be viewed as horizontal translations of one another.

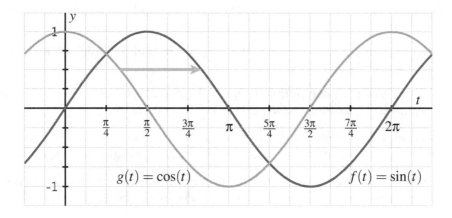

Figure 2.3.12: Graphs of the sine and cosine functions.

In particular, since the sine graph can be viewed as the cosine graph shifted $\frac{\pi}{2}$ units to the right, it follows that for any value of t,

$$\sin(t) = \cos(t - \frac{\pi}{2}).$$

Similarly, since the cosine graph can be viewed as the sine graph shifted left,

$$\cos(t) = \sin(t + \frac{\pi}{2}).$$

Because each of the two preceding equations hold for every value of t, they are often referred to as *identities*.

In light of the definitions of the sine and cosine functions, we can now view any point (x, y) on the unit circle as being of the form $(\cos(t), \sin(t))$, where t is the measure of the angle whose vertices are $(1, 0)$, $(0, 0)$, and (x, y). Note particularly that since $x^2 + y^2 = 1$, it is also true that $\cos^2(t) + \sin^2(t) = 1$. We call this fact the Fundamental Trigonometric Identity.

The Fundamental Trigonometric Identity.

For any real number t,
$$\cos^2(t) + \sin^2(t) = 1.$$

There are additional trends and patterns in the two functions' graphs that we explore further in the following activity.

Activity 2.3.3. Use Figure 2.3.12 to assist in answering the following questions.

a. Give an example of the largest interval you can find on which $f(t) = \sin(t)$ is decreasing.

b. Give an example of the largest interval you can find on which $f(t) = \sin(t)$ is decreasing and concave down.

c. Give an example of the largest interval you can find on which $g(t) = \cos(t)$ is increasing.

d. Give an example of the largest interval you can find on which $g(t) = \cos(t)$ is increasing and concave up.

e. Without doing any computation, on which interval is the average rate of change of $g(t) = \cos(t)$ greater: $[\pi, \pi + 0.1]$ or $[\frac{3\pi}{2}, \frac{3\pi}{2} + 0.1]$? Why?

f. In general, how would you characterize the locations on the sine and cosine graphs where the functions are increasing or decreasingly most rapidly?

g. Thinking from the perspective of the unit circle, for which quadrants of the x-y plane is $\cos(t)$ negative for an angle t that lies in that quadrant?

2.3.4 Using computing technology

We have established that we know the exact value of $\sin(t)$ and $\cos(t)$ for any of the t-values in Table 2.3.6, as well as for any such $t \pm 2j\pi$, where j is a whole number, due to the periodicity of the functions. But what if we want to know $\sin(1.35)$ or $\cos(\frac{\pi}{5})$ or values for other inputs not in the table?

Any standard computing device such as a scientific calculator, *Desmos, Geogebra,* or *WolframAlpha,* has the ability to evaluate the sine and cosine functions at any input we desire. Because the input is viewed as an angle, each computing device has the option to consider the angle in radians or degrees. *It is always essential that you are sure which type of input your device is expecting.* Our computational device of choice is *Desmos.* In *Desmos,* you can change the input type between radians and degrees by clicking the wrench icon in the upper right and choosing the desired units. Radian measure is the default.

It takes substantial and sophisticated mathematics to enable a computational device to evaluate the sine and cosine functions at any value we want; the algorithms involve an idea from calculus known as an infinite series. While your computational device is powerful, it's both helpful and important to understand the meaning of these values on the unit circle and to remember the special points for which we know the outputs of the sine and cosine functions exactly.

Activity 2.3.4. Answer the following questions exactly wherever possible. If you estimate a value, do so to at least 5 decimal places of accuracy.

a. The x coordinate of the point on the unit circle that lies in the third quadrant and whose y-coordinate is $y = -\frac{3}{4}$.

b. The y-coordinate of the point on the unit circle generated by a central angle in standard position that measures $t = 2$ radians.

c. The x-coordinate of the point on the unit circle generated by a central angle in standard position that measures $t = -3.05$ radians.

d. The value of $\cos(t)$ where t is an angle in Quadrant II that satisfies $\sin(t) = \frac{1}{2}$.

e. The value of $\sin(t)$ where t is an angle in Quadrant III for which $\cos(t) = -0.7$.

f. The average rate of change of $f(t) = \sin(t)$ on the intervals $[0.1, 0.2]$ and $[0.8, 0.9]$.

g. The average rate of change of $g(t) = \cos(t)$ on the intervals $[0.1, 0.2]$ and $[0.8, 0.9]$.

2.3.5 Summary

- The sine and cosine functions result from tracking the y- and x-coordinates of a point traversing the unit circle counterclockwise from $(1, 0)$. The value of $\sin(t)$ is the y-coordinate of a point that has traversed t units along the circle from $(1, 0)$ (or equivalently that corresponds to an angle of t radians), while the value of $\cos(t)$ is the x-coordinate of the same point.

- The sine and cosine functions are both periodic functions that share the same domain (the set of all real numbers), range (the interval $[-1, 1]$), midline ($y = 0$), amplitude ($a = 1$), and period ($P = 2\pi$). In addition, the sine function is horizontal shift of the cosine function by $\frac{\pi}{2}$ units to the right, so $\sin(t) = \cos(t - \frac{\pi}{2})$ for any value of t.

- If t corresponds to one of the special angles that we know on the unit circle (as in Figure 2.3.1), we can compute the values of $\sin(t)$ and $\cos(t)$ exactly. For other values of t, we can use a computational device to estimate the value of either function at a given input; when we do so, we must take care to know whether we are computing in terms of radians or degrees.

2.3.6 Exercises

1. Without using a calculator, find the exact value as fraction (not a decimal approximation).

$$\cos\left(\frac{4\pi}{3}\right)$$

2. Evaluate the following expressions.

    1. $\cos(-225°)$ = _____

    2. $\sin(135°)$ = _____

    3. $\cos(-150°)$ = _____

    4. $\sin(240°)$ = _____

    5. $\cos(225°)$ = _____

    6. $\sin(-180°)$ = _____

Remark: Your answer cannot contain trigonometric functions; it must be a fraction (not a decimal) and it may contain square roots (e.g., $\sqrt{2}$).

3. Determine whether each of the following expressions is Positive or Negative without using a calculator.

$\sin(70°)$ $\sin\left(\frac{15\pi}{16}\right)$

$\cos(138°)$ $\cos\left(\frac{20\pi}{21}\right)$

4. a) Write an expression (involving the variable a and h) for the slope of the line segment joining S and T in the figure below.

b) Evaluate your expression for $a = 1.6$ and $h = 0.01$. Round your answer to two decimal places.

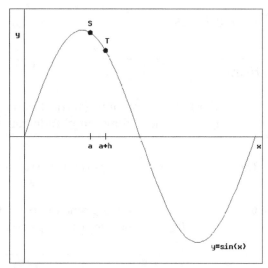

5. Without using a computational device, determine the exact value of each of the follow-ing quantities.

   a. $\sin(-\frac{11\pi}{4})$

   b. $\cos(\frac{29\pi}{6})$

   c. $\sin(47\pi)$

   d. $\cos(-113\pi)$

   e. t in quadrant III such that $\cos(t) = -\frac{\sqrt{3}}{2}$

   f. t in quadrant IV such that $\sin(t) = -\frac{\sqrt{3}}{2}$

6. We now know three different identities involving the sine and cosine functions: $\sin(t + \frac{\pi}{2}) = \cos(t)$, $\cos(t - \frac{\pi}{2}) = \sin(t)$, and $\cos^2(t) + \sin^2(t) = 1$. Following are several proposed identities. For each, your task is to decide whether the identity is true or false. If true, give a convincing argument for why it is true; if false, give an example of a t-value for which the equation fails to hold.

   a. $\cos(t + 2\pi) = \cos(t)$

   b. $\sin(t - \pi) = -\sin(t)$

   c. $\cos(t - \frac{3\pi}{2}) = \sin(t)$

   d. $\sin^2(t) = 1 - \cos^2(t)$

   e. $\sin(t) + \cos(t) = 1$

   f. $\sin(t) + \sin(\frac{\pi}{2}) = \cos(t)$

2.4 Sinusoidal Functions

Motivating Questions

- How do the three standard transformations (vertical translation, horizontal translation, and vertical scaling) affect the midline, amplitude, range, and period of sine and cosine curves?

- What algebraic transformation results in horizontal stretching or scaling of a function?

- How can we determine a formula involving sine or cosine that models any circular periodic function for which the midline, amplitude, period, and an anchor point are known?

Recall our work in Section 1.8, where we studied how the graph of the function g defined by $g(x) = af(x - b) + c$ is related to the graph of f, where a, b, and c are real numbers with $a \neq 0$. Because such transformations can shift and stretch a function, we are interested in understanding how we can use transformations of the sine and cosine functions to fit formulas to circular functions.

> **Preview Activity 2.4.1.** Let $f(t) = \cos(t)$. First, answer all of the questions below *without* using *Desmos*; then use *Desmos* to confirm your conjectures. For each prompt, describe the graphs of g and h as transformations of f and, in addition, state the amplitude, midline, and period of both g and h.
>
> a. $g(t) = 3\cos(t)$ and $h(t) = -\frac{1}{4}\cos(t)$
>
> b. $g(t) = \cos(t - \pi)$ and $h(t) = \cos\left(t + \frac{\pi}{2}\right)$
>
> c. $g(t) = \cos(t) + 4$ and $h(t) = \cos(t) - 2$
>
> d. $g(t) = 3\cos(t - \pi) + 4$ and $h(t) = -\frac{1}{4}\cos\left(t + \frac{\pi}{2}\right) - 2$

2.4.1 Shifts and vertical stretches of the sine and cosine functions

We know that the standard functions $f(t) = \sin(t)$ and $g(t) = \cos(t)$ are circular functions that each have midline $y = 0$, amplitude $a = 1$, period $p = 2\pi$, and range $[-1, 1]$. Our work in Preview Activity 2.4.1 suggests the following general principles.

> **Transformations of sine and cosine.**
>
> Given real numbers a, b, and c with $a \neq 0$, the functions
>
> $$k(t) = a\cos(t - b) + c \quad \text{and} \quad h(t) = a\sin(t - b) + c$$
>
> each represent a horizontal shift by b units to the right, followed by a vertical stretch by $|a|$ units (if $a < 0$, there is also a reflection across the x-axis), followed by a vertical

shift of c units, applied to the parent function ($\cos(t)$ or $\sin(t)$, respectively). The resulting circular functions have midline $y = c$, amplitude $|a|$, range $[c - |a|, c + |a|]$, and period $p = 2\pi$. In addition, the point $(b, a + c)$ lies on the graph of k and the point (b, c) lies on the graph of h.

In Figure 2.4.1, we see how the overall transformation $k(t) = a\cos(t - b) + c$ comes from executing a sequence of simpler ones. The original parent function $y = \cos(t)$ (in dark gray) is first shifted b units right to generate the light red graph of $y = \cos(t - b)$. In turn, that graph is then scaled vertically by a to generate the purple graph of $y = a\cos(t - b)$. Finally, the purple graph is shifted c units vertically to result in the final graph of $y = a\cos(t - b) + c$ in blue.

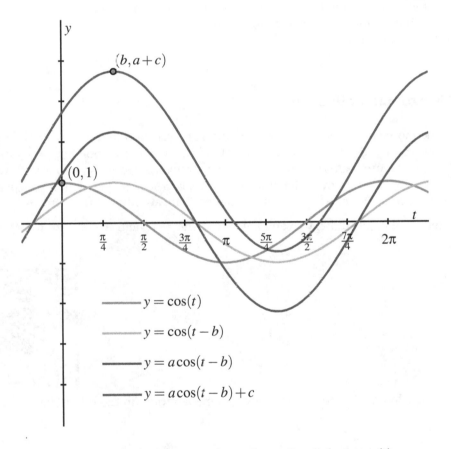

Figure 2.4.1: A sequence of transformations of $y = \cos(t)$.

It is often useful to follow one particular point through a sequence of transformations. In Figure 2.4.1, we see the red point that is located at $(0, 1)$ on the original function $y = \cos(t)$, as well as the point $(b, a + c)$ that is the corresponding point on $k(t) = a\cos(t - b) + c$ under the overall transformation. Note that the point $(b, a + c)$ results from the input, $t = b$, that makes the argument of the cosine function zero: $k(b) = a\cos(b - b) + c = a\cos(0) + c$.

While the sine and cosine functions extend infinitely in either direction, it's natural to think of the point $(0, 1)$ as the "starting point" of the cosine function, and similarly the point $(0, 0)$ as the starting point of the sine function. We will refer to the corresponding points $(b, a + c)$ and (b, c) on $k(t) = a \cos(t - b) + c$ and $h(t) = a \sin(t - b) + c$ as **anchor points**. Anchor points, along with other information about a circular function's amplitude, midline, and period help us to determine a formula for a function that fits a given situation.

> **Activity 2.4.2.** Consider a spring-mass system where a weight is resting on a friction-less table. We let $d(t)$ denote the distance from the wall (where the spring is attached) to the weight at time t in seconds and know that the weight oscillates periodically with a minimum value of $d(t) = 2$ feet and a maximum value of $d(t) = 7$ feet with a period of 2π. We also know that $d(0) = 4.5$ and $d\left(\frac{\pi}{2}\right) = 2$.
>
> Determine a formula for $d(t)$ in the form $d(t) = a \cos(t - b) + c$ or $d(t) = a \sin(t - b) + c$. Is it possible to find two different formulas that work? For any formula you find, identify the anchor point.

2.4.2 Horizontal scaling

There is one more very important transformation of a function that we've not yet explored. Given a function $y = f(x)$, we want to understand the related function $g(x) = f(kx)$, where k is a positive real number. The sine and cosine functions are ideal functions with which to explore these effects; moreover, this transformation is crucial for being able to use the sine and cosine functions to model phenomena that oscillate at different frequencies.

In the interactive Figure 2.4.2, we can explore the effect of the transformation $g(t) = f(kt)$, where $f(t) = \sin(t)$.

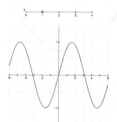 

Figure 2.4.2: Interactive horizontal scaling demonstration (in the HTML version only).

By experimenting with the slider, we gain an intuitive sense for how the value of k affects the graph of $h(t) = f(kt)$ in comparision to the graph of $f(t)$. When $k = 2$, we see that the graph of h is oscillating twice as fast as the graph of f since $h(t) = f(2t)$ completes two full cycles over an interval in which f completes one full cycle. In contrast, when $k = \frac{1}{2}$, the graph of h oscillates half as fast as the graph of f, as $h(t) = f(\frac{1}{2}t)$ completes only half of one cycle over an interval where $f(t)$ completes a full one.

We can also understand this from the perspective of function composition. To evaluate

$h(t) = f(2t)$, at a given value of t, we first multiply the input t by a factor of 2, and then evaluate the function f at the result. An important observation is that

$$h\left(\frac{1}{2}t\right) = f\left(2 \cdot \frac{1}{2}t\right) = f(t).$$

This tells us that the point $(\frac{1}{2}t, f(t))$ lies on the graph of h since an input of $\frac{1}{2}t$ in h results in the value $f(t)$. At the same time, the point $(t, f(t))$ lies on the graph of f. Thus we see that the correlation between points on the graphs of f and h (where $h(t) = f(2t)$) is

$$(t, f(t)) \rightarrow \left(\frac{1}{2}t, f(t)\right).$$

We can therefore think of the transformation $h(t) = f(2t)$ as achieving the output values of f twice as fast as the original function $f(t)$ does. Analogously, the transformation $h(t) = f(\frac{1}{2}t)$ will achieve the output values of f only half as quickly as the original function.

> **Horizontal scaling.**
>
> Given a function $y = f(t)$ and a real number $k > 0$, the transformed function $y = h(t) = f(kt)$ is a *horizontal stretch* of the graph of f. Every point $(t, f(t))$ on the graph of f gets stretched horizontally to the corresponding point $(\frac{1}{k}t, f(t))$ on the graph of h. If $0 < k < 1$, the graph of h is a stretch of f away from the y-axis by a factor of $\frac{1}{k}$; if $k > 1$, the graph of h is a compression of f toward the y-axis by a factor of $\frac{1}{k}$. The only point on the graph of f that is unchanged by the transformation is $(0, f(0))$.

While we will soon focus on horizontal stretches of the sine and cosine functions for the remainder of this section, it's important to note that horizontal scaling follows the same principles for any function we choose.

> **Activity 2.4.3.** Consider the functions f and g given in Figure 2.4.3 and Figure 2.4.4.
>
> a. On the same axes as the plot of $y = f(t)$, sketch the following graphs: $y = h(t) = f(\frac{1}{3}t)$ and $y = j(t) = r = f(4t)$. Be sure to label several points on each of f, h, and k with arrows to indicate their correspondence. In addition, write one sentence to explain the overall transformations that have resulted in h and j from f.
>
> b. On the same axes as the plot of $y = g(t)$, sketch the following graphs: $y = k(t) = g(2t)$ and $y = m(t) = g(\frac{1}{2}t)$. Be sure to label several points on each of g, k, and m with arrows to indicate their correspondence. In addition, write one sentence to explain the overall transformations that have resulted in k and m from g.

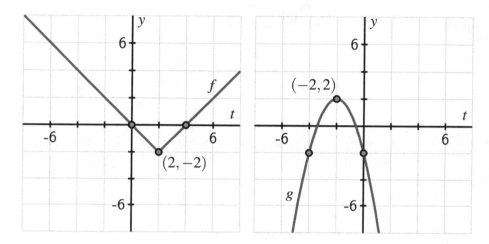

Figure 2.4.3: A parent function f.   **Figure 2.4.4:** A parent function g.

c. On the additional copies of the two figures below, sketch the graphs of the following transformed functions: $y = r(t) = 2f(\frac{1}{2}t)$ (at left) and $y = s(t) = \frac{1}{2}g(2t)$. As above, be sure to label several points on each graph and indicate their correspondence to points on the original parent function.

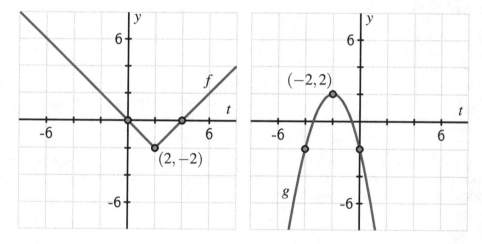

d. Describe in words how the function $y = r(t) = 2f(\frac{1}{2}t)$ is the result of two elementary transformations of $y = f(t)$. Does the order in which these transformations occur matter? Why or why not?

2.4.3 Circular functions with different periods

Because the circumference of the unit circle is 2π, the sine and cosine functions each have period 2π. Of course, as we think about using transformations of the sine and cosine functions to model different phenomena, it is apparent that we will need to generate functions with different periods than 2π. For instance, if a ferris wheel makes one revolution every 5 minutes, we'd want the period of the function that models the height of one car as a function of time to be $P = 5$. Horizontal scaling of functions enables us to generate circular functions with any period we desire.

We begin by considering two basic examples. First, let $f(t) = \sin(t)$ and $g(t) = f(2t) = \sin(2t)$. We know from our most recent work that this transformation results in a horizontal compression of the graph of $\sin(t)$ by a factor of $\frac{1}{2}$ toward the y-axis. If we plot the two functions on the same axes as seen in Figure 2.4.5, it becomes apparent how this transformation affects the period of f.

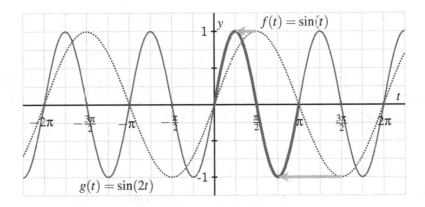

Figure 2.4.5: A plot of the parent function, $f(t) = \sin(t)$ (dashed, in gray), and the transformed function $g(t) = f(2t) = \sin(2t)$ (in blue).

From the graph, we see that $g(t) = \sin(2t)$ oscillates twice as frequently as $f(t) = \sin(t)$, and that g completes a full cycle on the interval $[0, \pi]$, which is half the length of the period of f. Thus, the "2" in $f(2t)$ causes the period of f to be $\frac{1}{2}$ as long; specifially, the period of g is $P = \frac{1}{2}(2\pi) = \pi$.

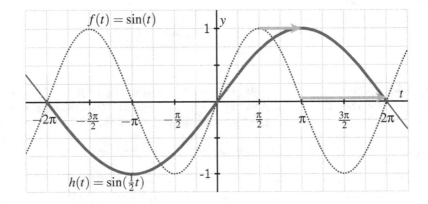

Figure 2.4.6: A plot of the parent function, $f(t) = \sin(t)$ (dashed, in gray), and the transformed function $h(t) = f(\frac{1}{2}t) = \sin(\frac{1}{2}t)$ (in blue).

On the other hand, if we let $h(t) = f(\frac{1}{2}t) = \sin(\frac{1}{2}t)$, the transformed graph h is stretched away from the y-axis by a factor of 2. This has the effect of doubling the period of f, so that the period of h is $P = 2 \cdot 2\pi = 4\pi$, as seen in Figure 2.4.6.

Our observations generalize for any positive constant $k > 0$. In the case where $k = 2$, we saw that the period of $g(t) = \sin(2t)$ is $P = \frac{1}{2} \cdot 2\pi$, whereas in the case where $k = \frac{1}{2}$, the period of $h(t) = \sin(\frac{1}{2}t)$ is $P = 2 \cdot 2\pi = \frac{1}{\frac{1}{2}} \cdot 2\pi$. Identical reasoning holds if we are instead working with the cosine function. In general, we can say the following.

> **The period of a circular function.**
>
> For any constant $k > 0$, the period of the functions $\sin(kt)$ and $\cos(kt)$ is
> $$P = \frac{2\pi}{k}.$$

Thus, if we know the k-value from the given function, we can deduce the period. If instead we know the desired period, we can determine k by the rule $k = \frac{2\pi}{P}$.

Activity 2.4.4. Determine the exact period, amplitude, and midline of each of the following functions. In addition, state the range of each function, any horizontal shift that has been introduced to the graph, and identify an anchor point. Make your conclusions without consulting *Desmos*, and then use the program to check your work.

a. $p(x) = \sin(10x) + 2$

b. $q(x) = -3\cos(0.25x) - 4$

c. $r(x) = 2\sin\left(\frac{\pi}{4}x\right) + 5$

d. $w(x) = 2\cos\left(\frac{\pi}{2}(x-3)\right) + 5$

e. $u(x) = -0.25\sin(3x - 6) + 5$

Activity 2.4.5. Consider a spring-mass system where the weight is hanging from the ceiling in such a way that the following is known: we let $d(t)$ denote the distance from the ceiling to the weight at time t in seconds and know that the weight oscillates periodically with a minimum value of $d(t) = 1.5$ and a maximum value of $d(t) = 4$, with a period of 3, and you know $d(0.5) = 2.75$ and $d(1.25) = 4$.

State the midline, amplitude, range, and an anchor point for the function, and hence determine a formula for $d(t)$ in the form $a\cos(k(t+b)) + c$ or $a\sin(k(t+b)) + c$. Show your work and thinking, and use *Desmos* appropriately to check that your formula generates the desired behavior.

2.4.4 Summary

- Given real numbers a, b, and c with $a \neq 0$, the functions

$$k(t) = a\cos(t - b) + c \text{ and } h(t) = a\sin(t - b) + c$$

each represent a horizontal shift by b units to the right, followed by a vertical stretch by $|a|$ units (with a reflection across the x-axis if $a < 0$), followed by a vertical shift of c units, applied to the parent function ($\cos(t)$ or $\sin(t)$, respectively). The resulting circular functions have midline $y = c$, amplitude $|a|$, range $[c - |a|, c + |a|]$, and period $p = 2\pi$. In addition, the anchor point $(b, a + c)$ lies on the graph of k and the anchor point (b, c) lies on the graph of h.

- Given a function f and a constant $k > 0$, the algebraic transformation $h(t) = f(kt)$ results in horizontal scaling of f by a factor of $\frac{1}{k}$. In particular, when $k > 1$, the graph of f is compressed toward the y-axis by a factor of $\frac{1}{k}$ to create the graph of h, while when $0 < k < 1$, the graph of f is stretched away from the y-axis by a factor of $\frac{1}{k}$ to create the graph of h.

- Given any circular periodic function for which the midline, amplitude, period, and an anchor point are known, we can find a corresponding formula for the function of the form

$$k(t) = a\cos(k(t - b)) + c \text{ or } h(t) = a\sin(k(t - b)) + c.$$

Each of these functions has midline $y = c$, amplitude $|a|$, and period $P = \frac{2\pi}{k}$. The point $(b, a + c)$ lies on k and the point (b, c) lies on h.

2.4.5 Exercises

1. Without a calculator, match each of the equations below to one of the graphs by placing the corresponding letter of the equation under the appropriate graph.

    A. $y = \sin(t) + 2$

    B. $y = \sin(t + 2)$

    C. $y = \sin(2t)$

    D. $y = 2\sin(t)$

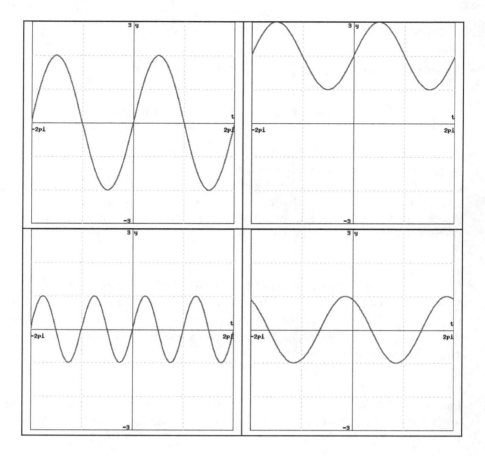

2. Find period, amplitude, and midline of the following function:

$$y = -6\cos(7\pi x + 2) - 9$$

3. Estimate the amplitude, midline, and period of the sinusoidal function graphed below:

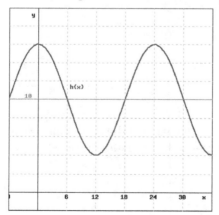

4. Find a possible formula for the trigonometric function graphed below.

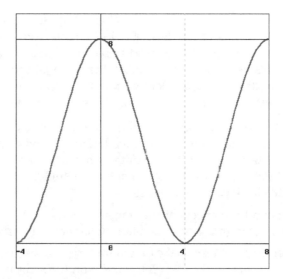

5. Find a formula for the trigonometric function graphed below. Use x as the independent variable in your formula.

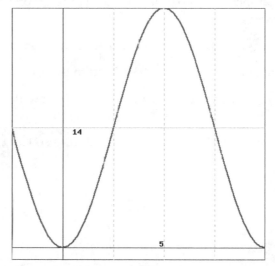

6. A ferris wheel is 40 meters in diameter and boarded at ground level. The wheel makes one full rotation every 8 minutes, and at time $t = 0$ you are at the 9 o'clock position and descending. Let $f(t)$ denote your height (in meters) above ground at t minutes. Find a formula for $f(t)$.

7. Find a possible formula for the circular function whose values are in the following table.

x	0	0.1	0.2	0.3	0.4	0.5	0.6	0.7	0.8	0.9	1.0
$g(x)$	2	2.6	3	3	2.6	2	1.4	1	1	1.4	2

Table 2.4.7: Data for a circular function.

Hint: Plot the points first; doing so in *Desmos* is ideal.

8. In 2018, on the summer solstice, June 21, Grand Rapids, MI, experiences 15 hours, 21 minutes, and 25 seconds of daylight. Said differently, on the 172nd day of the year, people on earth at the latitude of Grand Rapids experience 15.3569 hours of daylight. On the winter solstice, December 21, 2018, the same latitude has 9 hours, 0 minutes, and 31 seconds of daylight; equivalently, on day 355, there are 9.0086 hours of daylight. This data is essentially identical every year as the patterns of the earth's rotation repeat.

   Let t be the day of the year starting with $t = 0$ on December 31, 2017. In addition, let $s(t)$ be the number of hours of daylight on day t in Grand Rapids, MI. Find a formula for a circular function $s(t)$ that fits this data. What is the function's midline, amplitude, and period? What are you using as an anchor point? Explain fully, and then graph your function to check your conclusions.

9. We now understand the effects of the transformation $h(t) = f(kt)$ where $k > 0$ for a given function f. Our goal is to understand what happens when $k < 0$.

   a. We first consider the special case where $k = -1$. Let $f(t) = 2t - 1$, and let $g(t) = f(-1 \cdot t) = f(-t) = -2t - 1$. Plot f and g on the same coordinate axes. How are their graphs related to one another?

   b. Given any function p, how do you expect the graph of $y = q(t) = p(-t)$ to be related to the graph of p?

   c. How is the graph of $y = \sin(-3t)$ related to the graph of $y = \sin(t)$?

   d. How is the graph of $y = \cos(-3t)$ related to the graph of $y = \cos(t)$?

   e. Given any function p and a constant $k < 0$, how do you expect the graph of $y = q(t) = p(kt)$ to be related to the graph of p?

   f. How are $\sin(-t)$ and $\sin(t)$ related? How are $\cos(t)$ and $\cos(-t)$ related?

CHAPTER 3

Exponential and Logarithmic Functions

3.1 Exponential Growth and Decay

Motivating Questions

- What does it mean to say that a function is "exponential"?

- How much data do we need to know in order to determine the formula for an exponential function?

- Are there important trends that all exponential functions exhibit?

Linear functions have constant average rate of change and model many important phenomena. In other settings, it is natural for a quantity to change at a rate that is proportional to the amount of the quantity present. For instance, whether you put \$100 or \$100000 or any other amount in a mutual fund, the investment's value changes at a rate proportional the amount present. We often measure that rate in terms of the annual percentage rate of return.

Suppose that a certain mutual fund has a 10% annual return. If we invest \$100, after 1 year we still have the original \$100, plus we gain 10% of \$100, so

$$100 \xrightarrow{\text{year 1}} 100 + 0.1(100) = 1.1(100).$$

If we instead invested \$100000, after 1 year we again have the original \$100000, but now we gain 10% of \$100000, and thus

$$100000 \xrightarrow{\text{year 1}} 100000 + 0.1(100000) = 1.1(100000).$$

We therefore see that regardless of the amount of money originally invested, say P, the amount of money we have after 1 year is $1.1P$.

If we repeat our computations for the second year, we observe that

$$1.1(100) \xrightarrow{\text{year 2}} 1.1(100) + 0.1(1.1(100)) = 1.1(1.1(100)) = 1.1^2(100).$$

The ideas are identical with the larger dollar value, so

$$1.1(100000) \xrightarrow{\text{year 2}} 1.1(100000) + 0.1(1.1(100000)) = 1.1(1.1(100000)) = 1.1^2(100000),$$

and we see that if we invest P dollars, in 2 years our investment will grow to 1.1^2P.

Of course, in 3 years at 10%, the original investment P will have grown to 1.1^3P. Here we see a new kind of pattern developing: annual growth of 10% is leading to *powers* of the base 1.1, where the power to which we raise 1.1 corresponds to the number of years the investment has grown. We often call this phenomenon *exponential growth*.

Preview Activity 3.1.1. Suppose that at age 20 you have $20000 and you can choose between one of two ways to use the money: you can invest it in a mutual fund that will, on average, earn 8% interest annually, or you can purchase a new automobile that will, on average, depreciate 12% annually. Let's explore how the $20000 changes over time.

Let $I(t)$ denote the value of the $20000 after t years if it is invested in the mutual fund, and let $V(t)$ denote the value of the automobile t years after it is purchased.

    a. Determine $I(0)$, $I(1)$, $I(2)$, and $I(3)$.

    b. Note that if a quantity depreciates 12% annually, after a given year, 88% of the quantity remains. Compute $V(0)$, $V(1)$, $V(2)$, and $V(3)$.

    c. Based on the patterns in your computations in (a) and (b), determine formulas for $I(t)$ and $V(t)$.

    d. Use *Desmos* to define $I(t)$ and $V(t)$. Plot each function on the interval $0 \le t \le 20$ and record your results on the axes in Figure 3.1.1, being sure to label the scale on the axes. What trends do you observe in the graphs? How do $I(20)$ and $V(20)$ compare?

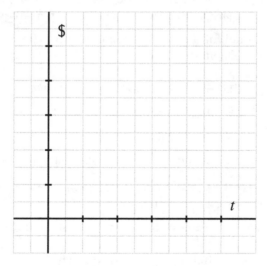

Figure 3.1.1: Blank axes for plotting I and V.

3.1.1 Exponential functions of form $f(t) = ab^t$

In Preview Activity 3.1.1, we encountered the functions $I(t)$ and $V(t)$ that had the same basic structure. Each can be written in the form $g(t) = ab^t$ where a and b are positive constants and $b \neq 1$. Based on our earlier work with transformations, we know that the constant a is a vertical scaling factor, and thus the main behavior of the function comes from b^t, which we call an "exponential function".

Definition 3.1.2 Let b be a real number such that $b > 0$ and $b \neq 1$. We call the function defined by

$$f(t) = b^t$$

an **exponential function** with **base** b. ◊

For an exponential function $f(t) = b^t$, we note that $f(0) = b^0 = 1$, so an exponential function of this form always passes through $(0, 1)$. In addition, because a positive number raised to any power is always positive (for instance, $2^{10} = 1096$ and $2^{-10} = \frac{1}{2^{10}} = \frac{1}{2096}$), the output of an exponential function is also always positive. In particular, $f(t) = b^t$ is never zero and thus has no x-intercepts.

Because we will be frequently interested in functions such as $I(t)$ and $V(t)$ with the form ab^t, we will also refer to functions of this form as "exponential", understanding that technically these are vertical stretches of exponential functions according to Definition 3.1.2. In Preview Activity 3.1.1, we found that $I(t) = 20000(1.08)^t$ and $V(t) = 20000(0.88)^t$. It is natural to call 1.08 the "growth factor" of I and similarly 0.88 the growth factor of V. In addition, we note that these values stem from the actual growth rates: 0.08 for I and -0.12 for V, the latter being negative because value is depreciating. In general, for a function of form $f(t) = ab^t$, we call b the **growth factor**. Moreover, if $b = 1+r$, we call r the **growth rate**. Whenever $b > 1$, we often say that the function f is exhibiting "exponential growth", whereas if $0 < b < 1$, we say f exhibits "exponential decay".

We explore the properties of functions of form $f(t) = ab^t$ further in Activity 3.1.2.

Activity 3.1.2. In *Desmos*, define the function $g(t) = ab^t$ and create sliders for both a and b when prompted. Click on the sliders to set the minimum value for each to 0.1 and the maximum value to 10. Note that for g to be an exponential function, we require $b \neq 1$, even though the slider for b will allow this value.

a. What is the domain of $g(t) = ab^t$?

b. What is the range of $g(t) = ab^t$?

c. What is the y-intercept of $g(t) = ab^t$?

d. How does changing the value of b affect the shape and behavior of the graph of $g(t) = ab^t$? Write several sentences to explain.

e. For what values of the growth factor b is the corresponding growth rate positive? For which b-values is the growth rate negative?

f. Consider the graphs of the exponential functions p and q provided in Figure 3.1.3. If $p(t) = ab^t$ and $q(t) = cd^t$, what can you say about the values a, b, c, and d

(beyond the fact that all are positive and $b \neq 1$ and $d \neq 1$)? For instance, can you say a certain value is larger than another? Or that one of the values is less than 1?

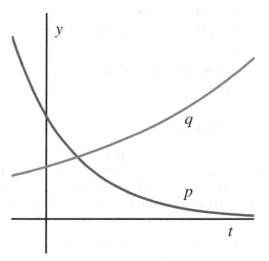

Figure 3.1.3: Graphs of exponential functions p and q.

3.1.2 Determining formulas for exponential functions

To better understand the roles that a and b play in an exponential function, let's compare exponential and linear functions. In Table 3.1.4 and Table 3.1.5, we see output for two different functions r and s that correspond to equally spaced input.

t	0	3	6	9
$r(t)$	12	10	8	6

t	0	3	6	9
$s(t)$	12	9	6.75	5.0625

Table 3.1.4: Data for the function r.   **Table 3.1.5:** Data for the function s.

In Table 3.1.4, we see a function that exhibits constant average rate of change since the change in output is always $\Delta r = -2$ for any change in input of $\Delta t = 3$. Said differently, r is a linear function with slope $m = -\frac{2}{3}$. Since its y-intercept is $(0, 12)$, the function's formula is $y = r(t) = 12 - \frac{2}{3}t$.

In contrast, the function s given by Table 3.1.5 does not exhibit constant average rate of change. Instead, another pattern is present. Observe that if we consider the ratios of consecutive outputs in the table, we see that

$$\frac{9}{12} = \frac{3}{4}, \frac{6.75}{9} = 0.75 = \frac{3}{4}, \text{ and } \frac{5.0625}{6.75} = 0.75 = \frac{3}{4}.$$

So, where the *differences* in the outputs in Table 3.1.4 are constant, the *ratios* in the outputs in Table 3.1.5 are constant. The latter is a hallmark of exponential functions and may be used to help us determine the formula of a function for which we have certain information.

If we know that a certain function is linear, it suffices to know two points that lie on the line to determine the function's formula. It turns out that exponential functions are similar: knowing two points on the graph of a function known to be exponential is enough information to determine the function's formula. In the following example, we show how knowing two values of an exponential function enables us to find both a and b exactly.

Example 3.1.6 Suppose that p is an exponential function and we know that $p(2) = 11$ and $p(5) = 18$. Determine the exact values of a and b for which $p(t) = ab^t$.

Solution.   Since we know that $p(t) = ab^t$, the two data points give us two equations in the unknowns a and b. First, using $t = 2$,

$$ab^2 = 11, \qquad\qquad (3.1.1)$$

and using $t = 5$ we also have

$$ab^5 = 18. \qquad\qquad (3.1.2)$$

Because we know that the quotient of outputs of an exponential function corresponding to equally-spaced inputs must be constant, we thus naturally consider the quotient $\frac{18}{11}$. Using Equation (3.1.1) and Equation (3.1.2), it follows that

$$\frac{18}{11} = \frac{ab^5}{ab^2}.$$

Simplifying the fraction on the right, we see that $\frac{18}{11} = b^3$. Solving for b, we find that $b = \sqrt[3]{\frac{18}{11}}$ is the exact value of b. Substituting this value for b in Equation (3.1.1), it then follows that $a\left(\sqrt[3]{\frac{18}{11}}\right)^2 = 11$, so $a = \frac{11}{\left(\frac{18}{11}\right)^{2/3}}$. Therefore,

$$p(t) = \frac{11}{\left(\frac{18}{11}\right)^{2/3}} \left(\sqrt[3]{\frac{18}{11}}\right)^t \approx 7.9215 \cdot 1.1784^t,$$

and a plot of $y = p(t)$ confirms that the function indeed passes through $(2, 11)$ and $(5, 18)$ as shown in Figure 3.1.7.

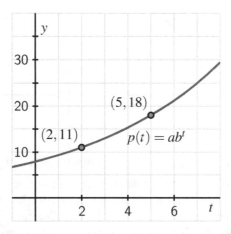

Figure 3.1.7: Plot of $p(t) = ab^t$ that passes through $(2, 11)$ and $(5, 18)$.

□

Activity 3.1.3. The value of an automobile is depreciating. When the car is 3 years old, its value is $12500; when the car is 7 years old, its value is $6500.

    a. Suppose the car's value t years after its purchase is given by the function $V(t)$ and that V is exponential with form $V(t) = ab^t$, what are the exact values of a and b?

    b. Using the exponential model determined in (a), determine the purchase value of the car and estimate when the car will be worth less than $1000.

    c. Suppose instead that the car's value is modeled by a linear function L and satisfies the values stated at the outset of this activity. Find a formula for $L(t)$ and determine both the purchase value of the car and when the car will be worth $1000.

    d. Which model do you think is more realistic? Why?

3.1.3 Trends in the behavior of exponential functions

Recall that a function is increasing on an interval if its value always increasing as we move from left to right. Similarly, a function is decreasing on an interval provided that its value always decreases as we move from left to right.

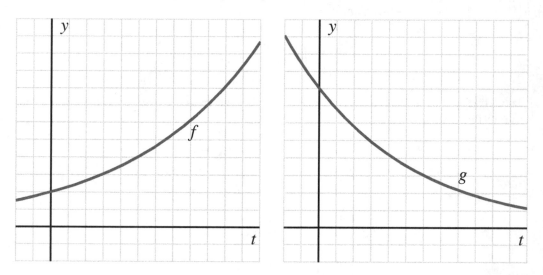

Figure 3.1.8: The exponential function f.   **Figure 3.1.9:** The exponential function g.

If we consider an exponential function f with a growth factor $b > 1$, such as the function pictured in Figure 3.1.8, then the function is always increasing because higher powers of b are greater than lesser powers (for example, $(1.2)^3 > (1.2)^2$). On the other hand, if $0 < b < 1$, then the exponential function will be decreasing because higher powers of positive numbers less than 1 get smaller (e.g., $(0.9)^3 < (0.9)^2$), as seen for the exponential function in Figure 3.1.9.

An additional trend is apparent in the graphs in Figure 3.1.8 and Figure 3.1.9. Each graph bends upward and is therefore concave up. We can better understand why this is so by considering the average rate of change of both f and g on consecutive intervals of the same width. We choose adjacent intervals of length 1 and note particularly that as we compute the average rate of change of each function on such intervals,

$$AV_{[t,t+1]} = \frac{f(t+1) - f(t)}{t+1-1} = f(t+1) - f(t).$$

Thus, these average rates of change are also measuring the total change in the function across an interval that is 1-unit wide. We now assume that $f(t) = 2(1.25)^t$ and $g(t) = 8(0.75)^t$ and compute the rate of change of each function on several consecutive intervals.

t	$f(t)$	$AV_{[t,t+1]}$
0	2	0.5
1	2.5	0.625
2	3.125	0.78215
3	3.90625	0.97656

t	$g(t)$	$AV_{[t,t+1]}$
0	8	−2
1	6	−1.5
2	4.5	−1.125
3	3.375	−0.84375

Table 3.1.10: The average rate of change of $f(t) = 2(1.25)^t$.

Table 3.1.11: The average rate of change of $g(t) = 8(0.75)^t$.

From the data in Table 3.1.10, we see that the average rate of change is increasing as we increase the value of t. We naturally say that f appears to be "increasing at an increasing rate". For the function g, we first notice that its average rate of change is always negative, but also that the average rate of change gets *less negative* as we increase the value of t. Said differently, the average rate of change of g is also increasing as we increase the value of t. Since g is always decreasing but its average rate of change is increasing, we say that g appears to be "decreasing at an increasing rate". These trends hold for exponential functions generally[1] according to the following conditions.

> **Trends in exponential function behavior.**
>
> For an exponential function of the form $f(t) = ab^t$ where a and b are both positive with $b \neq 1$,
>
> - if $b > 1$, then f is always increasing and always increases at an increasing rate;
>
> - if $0 < b < 1$, then f is always decreasing and always decreases at an increasing rate.

Observe how a function's average rate of change helps us classify the function's behavior on an interval: whether the average rate of change is always positive or always negative on the interval enables us to say if the function is always increasing or always decreasing, and then how the average rate of change itself changes enables us to potentially say *how* the function is increasing or decreasing through phrases such as "decreasing at an increasing rate".

---

[1]It takes calculus to justify this claim fully and rigorously.

Activity 3.1.4. For each of the following prompts, give an example of a function that satisfies the stated characteristics by both providing a formula and sketching a graph.

    a. A function p that is always decreasing and decreases at a constant rate.

    b. A function q that is always increasing and increases at an increasing rate.

    c. A function r that is always increasing for $t < 2$, always decreasing for $t > 2$, and is always changing at a decreasing rate.

    d. A function s that is always increasing and increases at a decreasing rate. (Hint: to find a formula, think about how you might use a transformation of a familiar function.)

    e. A function u that is always decreasing and decreases at a decreasing rate.

3.1.4 Summary

- We say that a function is exponential whenever its algebraic form is $f(t) = ab^t$ for some positive constants a and b where $b \neq 1$. (Technically, the formal definition of an exponential function is one of form $f(t) = b^t$, but in our everyday usage of the term "exponential" we include vertical stretches of these functions and thus allow a to be any positive constant, not just $a = 1$.)

- To determine the formula for an exponential function of form $f(t) = ab^t$, we need to know two pieces of information. Typically this information is presented in one of two ways.

    ○ If we know the amount, a, of a quantity at time $t = 0$ and the rate, r, at which the quantity grows or decays per unit time, then it follows $f(t) = a(1 + r)^t$. In this setting, r is often given as a percentage that we convert to a decimal (e.g., if the quantity grows at a rate of 7% per year, we set $r = 0.07$, so $b = 1.07$).

    ○ If we know any two points on the exponential function's graph, then we can set up a system of two equations in two unknowns and solve for both a and b exactly. In this situation, it is useful to consider the quotient of the two known outputs, as demonstrated in Example 3.1.6.

- Exponential functions of the form $f(t) = ab^t$ (where a and b are both positive and $b \neq 1$) exhibit the following important characteristics:

    ○ The domain of any exponential function is the set of all real numbers and the range of any exponential function is the set of all positive real numbers.

    ○ The y-intercept of the exponential function $f(t) = ab^t$ is $(0, a)$ and the function has no x-intercepts.

    ○ If $b > 1$, then the exponential function is always increasing and always increases at an increasing rate. If $0 < b < 1$, then the exponential function is always decreasing and always decreases at an increasing rate.

3.1.5 Exercises

1. Suppose $Q = 30.8(0.751)^t$. Give the starting value a, the growth *factor* b, and the growth *rate* r if $Q = a \cdot b^t = a(1 + r)^t$.

2. Find a formula for $P = f(t)$, the size of the population that begins in year $t = 0$ with 2090 members and decreases at a 3.7 % annual rate. Assume that time is measured in years.

3. (a) The annual inflation rate is 3.8% per year. If a movie ticket costs \$9.00 today, find a formula for p, the price of a movie ticket t years from today, assuming that movie tickets keep up with inflation.

   (b) According to your formula, how much will a movie ticket cost in 30 years?

4. In the year 2003, a total of 7.2 million passengers took a cruise vacation. The global cruise industry has been growing at 9% per year for the last decade. Assume that this growth rate continues.

   (a) Write a formula for to approximate the number, N, of cruise passengers (in millions) t years after 2003.

   (b) How many cruise passengers (in millions) are predicted in the year 2011?

   (c) How many cruise passengers (in millions) were there in the year 2000?

5. The populations, P, of six towns with time t in years are given by

1	$P = 800(0.78)^t$
2	$P = 900(1.06)^t$
3	$P = 1600(0.96)^t$
4	$P = 1400(1.187)^t$
5	$P = 500(1.14)^t$
6	$P = 2800(0.8)^t$

Answer the following questions regarding the populations of the six towns above.

(a) Which of the towns are growing?

(b) Which of the towns are shrinking?

(c) Which town is growing the fastest?

What is the annual percentage growth RATE of that town?

(d) Which town is shrinking the fastest?

What is the annual percentage decay RATE of that town?

(e) Which town has the largest initial population?

(f) Which town has the smallest initial population?

6. (a) Determine whether function whose values are given in the table below could be linear or exponential.

$x =$	0	1	2	3	4
$h(x) =$	14	8	2	-4	-10

Find a possible formula for this function.

(b) Determine whether function whose values are given in the table below could be linear or exponential.

$x =$	0	1	2	3	4
$i(x) =$	14	12.6	11.34	10.206	9.1854

Find a possible formula for this function.

7. A population has size 8000 at time $t = 0$, with t in years.

(a) If the population decreases by 125 people per year, find a formula for the population, P, at time t.

(b) If the population decreases by 6% per year, find a formula for the population, P, at time t.

8. Grinnell Glacier in Glacier National Park in Montana covered about 142 acres in 2007 and was found to be shrinking at about 4.4% per year.[2]

  a. Let $G(t)$ denote the area of Grinnell Glacier in acres in year t, where t is the number of years since 2007. Find a formula for $G(t)$ and define the function in *Desmos*.

  b. How many acres of ice were in the glacier in 1997? In 2012? What does the model predict for 2022?

  c. How many total acres of ice were lost from 2007 to 2012?

  d. What was the average rate of change of G from 2007 to 2012? Write a sentence to explain the meaning of this number and include units on your answer. In addition, how does this compare to the average rate of change of G from 2012 to 2017?

  e. How would you you describe the overall behavior of G, and thus what is happening to the Grinnell Glacier?

9. Consider the exponential function f whose graph is given by Figure 3.1.12. Note that f passes through the two noted points exactly.

  a. Determine the values of a and b exactly.

  b. Determine the average rate of change of f on the intervals $[2, 7]$ and $[7, 12]$. Which average rate is greater?

  c. Find the equation of the linear function L that passes through the points $(2, 20)$ and $(7, 5)$.

---

[2]See Exercise 34 on p. 146 of Connally's *Functions Modeling Change*.

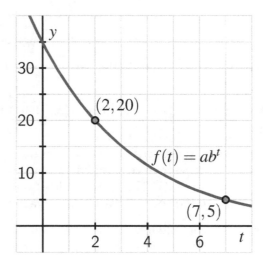

Figure 3.1.12: A plot of the exponential function f.

    d. Which average rate of change is greater? The average rate of change of f on $[0,2]$ or the average rate of change of L on $[0,2]$?

10. A cup of hot coffee is brought outside on a cold winter morning in Winnipeg, Manitoba, where the surrounding temperature is 0 degrees Fahrenheit. A temperature probe records the coffee's temperature (in degrees Fahrenheit) every minute and generates the data shown in Table 3.1.13.

t	0	2	4	6	8	10
$F(t)$	175	129.64	96.04	71.15	52.71	39.05

Table 3.1.13: The temperature, F, of the coffee at time t.

    a. Assume that the data in the table represents the overall trend of the behavior of F. Is F linear, exponential, or neither? Why?

    b. Is it possible to determine an exact formula for F? If yes, do so and justify your formula; if not, explain why not.

    c. What is the average rate of change of F on $[4,6]$? Write a sentence that explains the practical meaning of this value in the context of the overall exercise.

    d. How do you think the data would appear if instead of being in a regular coffee cup, the coffee was contained in an insulated mug?

11. The amount (in milligrams) of a drug in a person's body following one dose is given by an exponential decay function. Let $A(t)$ denote the amount of drug in the body at time t in hours after the dose was taken. In addition, suppose you know that $A(3) = 22.7$ and $A(6) = 15.2$.

    a. Find a formula for A in the form $A(t) = ab^t$, where you determine the values of a

and *b* exactly.

b. What is the size of the initial dose the person was given?

c. How much of the drug remains in the person's body 8 hours after the dose was taken?

d. Estimate how long it will take until there is less than 1 mg of the drug remaining in the body.

e. Compute the average rate of change of *A* on the intervals [3, 5], [5, 7], and [7, 9]. Write at least one careful sentence to explain the meaning of the values you found, including appropriate units. Then write at least one additional sentence to explain any overall trend(s) you observe in the average rate of change.

f. Plot *A*(*t*) on an appropriate interval and label important points and features of the graph to highlight graphical interpretations of your answers in (b), (c), (d), and (e).

3.2 Modeling with exponential functions

Motivating Questions

- What can we say about the behavior of an exponential function as the input gets larger and larger?

- How do vertical stretches and shifts of an exponontial function affect its behavior?

- Why is the temperature of a cooling or warming object modeled by a function of the form $F(t) = ab^t + c$?

If a quantity changes so that its growth or decay occurs at a constant percentage rate with respect to time, the function is exponential. This is because if the growth or decay rate is r, the total amount of the quantity at time t is given by $A(t) = a(1 + r)^t$, where a is the amount present at time $t = 0$. Many different natural quantities change according to exponential models: money growth through compounding interest, the growth of a population of cells, and the decay of radioactive elements.

A related situation arises when an object's temperature changes in response to its surroundings. For instance, if we have a cup of coffee at an initial temperature of 186° Fahrenheit and the cup is placed in a room where the surrounding temperature is 71°, our intuition and experience tell us that over time the coffee will cool and eventually tend to the 71° temperature of the surroundings. From an experiment [1] with an actual temperature probe, we have the data in Table 3.2.1 that is plotted in Figure 3.2.2.

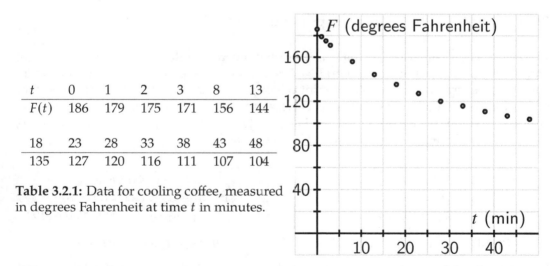

t	0	1	2	3	8	13
$F(t)$	186	179	175	171	156	144

18	23	28	33	38	43	48
135	127	120	116	111	107	104

Table 3.2.1: Data for cooling coffee, measured in degrees Fahrenheit at time t in minutes.

Figure 3.2.2: A plot of the data in Table 3.2.1.

In one sense, the data looks exponential: the points appear to lie on a curve that is always decreasing and decreasing at an increasing rate. However, we know that the function can't

---

[1]See http://gvsu.edu/s/0SB for this data.

have the form $f(t) = ab^t$ because such a function's range is the set of all positive real numbers, and it's impossible for the coffee's temperature to fall below room temperature (71°). It is natural to wonder if a function of the form $g(t) = ab^t + c$ will work. Thus, in order to find a function that fits the data in a situation such as Figure 3.2.2, we begin by investigating and understanding the roles of a, b, and c in the behavior of $g(t) = ab^t + c$.

> **Preview Activity 3.2.1.** In *Desmos*, define $g(t) = ab^t + c$ and accept the prompt for sliders for both a and b. Edit the sliders so that a has values from $a = 5$ to $a = 50$, b has values from $b = 0.7$ to $b = 1.3$, and c has values from $c = -5$ to $c = 5$ (also with a step-size of 0.01). In addition, in *Desmos* let $P = (0, g(0))$ and check the box to show the label. Finally, zoom out so that the window shows an interval of t-values from $-30 \le t \le 30$.
>
> a. Set $b = 1.1$ and explore the effects of changing the values of a and c. Write several sentences to summarize your observations.
>
> b. Follow the directions for (a) again, this time with $b = 0.9$
>
> c. Set $a = 5$ and $c = 4$. Explore the effects of changing the value of b; be sure to include values of b both less than and greater than 1. Write several sentences to summarize your observations.
>
> d. When $0 < b < 1$, what happens to the graph of g when we consider positive t-values that get larger and larger?

3.2.1 Long-term behavior of exponential functions

We have already established that any exponential function of the form $f(t) = ab^t$ where a and b are positive real numbers with $b \ne 1$ is always concave up and is either always increasing or always decreasing. We next introduce precise language to describe the behavior of an exponential function's value as t gets bigger and bigger. To start, let's consider the two basic exponential functions $p(t) = 2^t$ and $q(t) = (\frac{1}{2})^t$ and their respective values at $t = 10$, $t = 20$, and $t = 30$, as displayed in Table 3.2.3 and Table 3.2.4.

t	$p(t)$
10	$2^{10} = 1026$
20	$2^{20} = 1048576$
30	$2^{30} = 1073741824$

t	$q(t)$
10	$(\frac{1}{2})^{10} = \frac{1}{1026} \approx 0.00097656$
20	$(\frac{1}{2})^{20} = \frac{1}{1048576} \approx 0.00000095367$
30	$(\frac{1}{2})^{30} = \frac{1}{1073741824} \approx 0.00000000093192$

Table 3.2.3: Data for $p(t) = 2^t$.   **Table 3.2.4:** Data for $q(t) = (\frac{1}{2})^t$.

For the increasing function $p(t) = 2^t$, we see that the output of the function gets very large very quickly. In addition, there is no upper bound to how large the function can be. Indeed, we can make the value of $p(t)$ as large as we'd like by taking t sufficiently big. We thus say that as t increases, $p(t)$ **increases without bound**.

For the decreasing function $q(t) = (\frac{1}{2})^t$, we see that the output $q(t)$ is always positive but getting closer and closer to 0. Indeed, becasue we can make 2^t as large as we like, it follows that we can make its reciprocal $\frac{1}{2^t} = (\frac{1}{2})^t$ as small as we'd like. We thus say that as t increases, $q(t)$ **approaches** 0.

To represent these two common phenomena with exponential functions—the value increasing without bound or the value approaching 0—we will use shorthand notation. First, it is natural to write "$q(t) \to 0$" as t increases without bound. Moreover, since we have the notion of the infinite to represent quantities without bound, we use the symbol for infinity and arrow notation (∞) and write "$p(t) \to \infty$" as t increases without bound in order to indicate that $p(t)$ increases without bound.

In Preview Activity 3.2.1, we saw how the value of b affects the steepness of the graph of $f(t) = ab^t$, as well as how all graphs with $b > 1$ have the similar increasing behavior, and all graphs with $0 < b < 1$ have similar decreasing behavior. For instance, by taking t sufficiently large, we can make $(1.01)^t$ as large as we want; it just takes much larger t to make $(1.01)^t$ big in comparison to 2^t. In the same way, we can make $(0.99)^t$ as close to 0 as we wish by taking t sufficiently big, even though it takes longer for $(0.99)^t$ to get close to 0 in comparison to $(\frac{1}{2})^t$. For an arbitrary choice of b, we can say the following.

Long-term behavior of exponential functions.

Let $f(t) = b^t$ with $b > 0$ and $b \neq 1$.

- If $0 < b < 1$, then $b^t \to 0$ as $t \to \infty$. We read this notation as "b^t tends to 0 as t increases without bound."

- If $b > 1$, then $b^t \to \infty$ as $t \to \infty$. We read this notation as "b^t increases without bound as t increases without bound."

In addition, we make a key observation about the use of exponents. For the function $q(t) = (\frac{1}{2})^t$, there are three equivalent ways we may write the function:

$$\left(\frac{1}{2}\right)^t = \frac{1}{2^t} = 2^{-t}.$$

In our work with transformations involving horizontal scaling in Exercise 2.4.5.9, we saw that the graph of $y = h(-t)$ is the reflection of the graph of $y = h(t)$ across the y-axis. Therefore, we can say that the graphs of $p(t) = 2^t$ and $q(t) = (\frac{1}{2})^t = 2^{-t}$ are reflections of one another in the y-axis since $p(-t) = 2^{-t} = q(t)$. We see this fact verified in Figure 3.2.5. Similar observations hold for the relationship between the graphs of b^t and $\frac{1}{b^t} = b^{-t}$ for any positive $b \neq 1$.

3.2.2 The role of c in $g(t) = ab^t + c$

The function $g(t) = ab^t + c$ is a vertical translation of the function $f(t) = ab^t$. We now have extensive understanding of the behavior of $f(t)$ and how that behavior depends on a and b. Since a vertical translation by c does not change the shape of any graph, we expect that g will exhibit very similar behavior to f. Indeed, we can compare the two functions' graphs as shown in Figure 3.2.6 and Figure 3.2.7 and then make the following general observations.

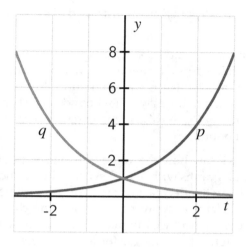

Figure 3.2.5: Plots of $p(t) = 2^t$ and $q(t) = 2^{-t}$.

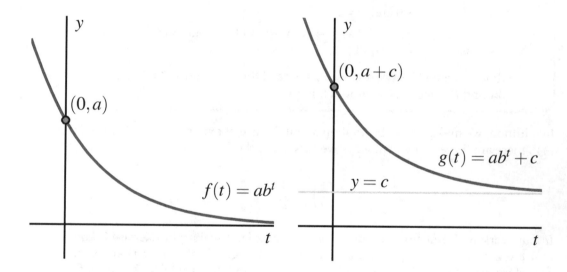

Figure 3.2.6: Plot of $f(t) = ab^t$.

Figure 3.2.7: Plot of $g(t) = ab^t + c$.

Behavior of vertically shifted exponential functions.

Let $g(t) = ab^t + c$ with $a > 0$, $b > 0$ and $b \neq 1$, and c any real number.

- If $0 < b < 1$, then $g(t) = ab^t + c \to c$ as $t \to \infty$. The function g is always decreasing, always concave up, and has y-intercept $(0, a + c)$. The range of the function is all real numbers greater than c.

- If $b > 1$, then $g(t) = ab^t + c \to \infty$ as $t \to \infty$. The function g is always increasing, always concave up, and has y-intercept $(0, a + c)$. The range of the function is all real numbers greater than c.

It is also possible to have $a < 0$. In this situation, because $g(t) = ab^t$ is both a reflection of $f(t) = b^t$ across the x-axis and a vertical stretch by $|a|$, the function g is always concave down. If $0 < b < 1$ so that f is always decreasing, then g is always increasing; if instead $b > 1$ so f is increasing, then g is decreasing. Moreover, instead of the range of the function g having a lower bound as when $a > 0$, in this setting the range of g has an upper bound. These ideas are explored further in Activity 3.2.2.

It's an important skill to be able to look at an exponential function of the form $g(t) = ab^t + c$ and form an accurate mental picture of the graph's main features in light of the values of a, b, and c.

> **Activity 3.2.2.** For each of the following functions, *without* using graphing technology, determine whether the function is
>
>   i. always increasing or always decreasing;
>
>   ii. always concave up or always concave down; and
>
> iii. increasing without bound, decreasing without bound, or increasing/decreasing toward a finite value.
>
> In addition, state the y-intercept and the range of the function. For each function, write a sentence that explains your thinking and sketch a rough graph of how the function appears.
>
>   a. $p(t) = 4372(1.000235)^t + 92856$ d. $s(t) = -17398(0.85234)^t + 19411$
>
>   b. $q(t) = 27931(0.97231)^t + 549786$ e. $u(t) = -7522(1.03817)^t$
>
>   c. $r(t) = -17398(0.85234)^t$ f. $v(t) = -7522(1.03817)^t + 6731$

3.2.3 Modeling temperature data

Newton's Law of Cooling states that the rate that an object warms or cools occurs in direct proportion to the difference between its own temperature and the temperature of its surroundings. If we return to the coffee temperature data in Table 3.2.1 and recall that the room temperature in that experiment was $71°$, we can see how to use a transformed exponential function to model the data. In Table 3.2.8, we add a row of information to the table where we compute $F(t) - 71$ to subtract the room temperature from each reading.

t	0	1	2	3	8	13	18	23	28	33	38	43	48
$F(t)$		186	179	175	171	156	144	135	127	120	116	111	107	104
$f(t) = F(t) - 71$	115	108	104	100	85	73	64	56	49	45	40	36	33

Table 3.2.8: Data for cooling coffee, measured in degrees Fahrenheit at time t in minutes, plus shifted to account for room temperature.

The data in the bottom row of Table 3.2.8 appears exponential, and if we test the data by computing the quotients of output values that correspond to equally-spaced input, we see a nearly constant ratio. In particular,

$$\frac{73}{85} \approx 0.86, \ \frac{64}{73} \approx 0.88, \ \frac{56}{64} \approx 0.88, \ \frac{49}{56} \approx 0.88, \ \frac{45}{49} \approx 0.92, \text{and} \frac{40}{45} \approx 0.89.$$

Of course, there is some measurement error in the data (plus it is only recorded to accuracy of whole degrees), so these computations provide convincing evidence that the underlying function is exponential. In addition, we expect that if the data continued in the bottom row of Table 3.2.8, the values would approach 0 because $F(t)$ will approach 71.

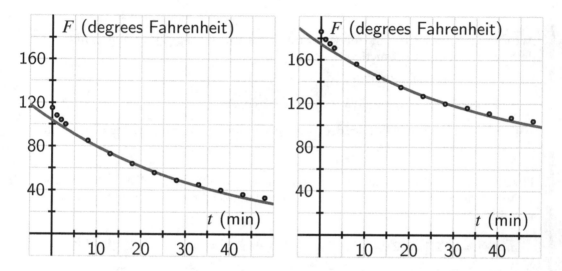

Figure 3.2.9: Plot of $f(t) = 103.503(0.974)^t$.   **Figure 3.2.10:** Plot of $F(t) = 103.503(0.974)^t + 71$.

If we choose two of the data points, say $(18, 64)$ and $(23, 56)$, and assume that $f(t) = ab^t$, we can determine the values of a and b. Doing so, it turns out that $a \approx 103.503$ and $b \approx 0.974$, so $f(t) = 103.503(0.974)^t$. Since $f(t) = F(t) - 71$, we see that $F(t) = f(t) + 71$, so $F(t) = 103.503(0.974)^t + 71$. Plotting f against the shifted data and F along with the original data in Figure 3.2.9 and Figure 3.2.10, we see that the curves go exactly through the points where $t = 18$ and $t = 23$ as expected, but also that the function provides a reasonable model for the observed behavior at any time t. If our data was even more accurate, we would expect that the curve's fit would be even better.

Our preceding work with the coffee data can be done similarly with data for any cooling or warming object whose temperature initially differs from its surroundings. Indeed, it is possible to show that Newton's Law of Cooling implies that the object's temperature is given by a function of the form $F(t) = ab^t + c$.

> **Activity 3.2.3.** A can of soda (at room temperature) is placed in a refrigerator at time $t = 0$ (in minutes) and its temperature, $F(t)$, in degrees Fahrenheit, is computed at regular intervals. Based on the data, a model is formulated for the object's temperature, given by
>
> $$F(t) = 42 + 30(0.95)^t.$$

a. Consider the simpler (parent) function $p(t) = (0.95)^t$. How do you expect the graph of this function to appear? How will it behave as time increases? Without using graphing technology, sketch a rough graph of p and write a sentence of explanation.

b. For the slightly more complicated function $r(t) = 30(0.95)^t$, how do you expect this function to look in comparison to p? What is the long-range behavior of this function as t increases? Without using graphing technology, sketch a rough graph of r and write a sentence of explanation.

c. Finally, how do you expect the graph of $F(t) = 42 + 30(0.95)^t$ to appear? Why? First sketch a rough graph without graphing technology, and then use technology to check your thinking and report an accurate, labeled graph on the axes provided in Figure 3.2.11.

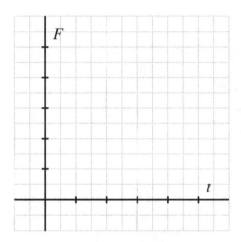

Figure 3.2.11: Axes for plotting F.

d. What is the temperature of the refrigerator? What is the room temperature of the surroundings outside the refrigerator? Why?

e. Determine the average rate of change of F on the intervals $[10, 20]$, $[20, 30]$, and $[30, 40]$. Write at least two careful sentences that explain the meaning of the values you found, including units, and discuss any overall trend in how the average rate of change is changing.

Activity 3.2.4. A potato initially at room temperature ($68°$) is placed in an oven (at $350°$) at time $t = 0$. It is known that the potato's temperature at time t is given by the function $F(t) = a - b(0.98)^t$ for some positive constants a and b, where F is measured in degrees Fahrenheit and t is time in minutes.

   a. What is the numerical value of $F(0)$? What does this tell you about the value of $a - b$?

   b. Based on the context of the problem, what should be the long-range behavior of the function $F(t)$? Use this fact along with the behavior of $(0.98)^t$ to determine the value of a. Write a sentence to explain your thinking.

   c. What is the value of b? Why?

   d. Check your work above by plotting the function F using graphing technology in an appropriate window. Record your results on the axes provided in Figure 3.2.12, labeling the scale on the axes. Then, use the graph to estimate the time at which the potato's temperature reaches 325 degrees.

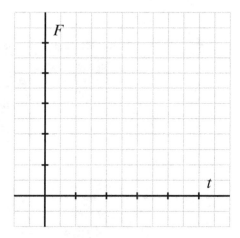

Figure 3.2.12: Axes for plotting F.

   e. How can we view the function $F(t) = a - b(0.98)^t$ as a transformation of the parent function $f(t) = (0.98)^t$? Explain.

3.2.4 Summary

- For an exponential function of the form $f(t) = b^t$, the function either approaches zero or grows without bound as the input gets larger and larger. In particular, if $0 < b < 1$, then $f(t) = b^t \to 0$ as $t \to \infty$, while if $b > 1$, then $f(t) = b^t \to \infty$ as $t \to \infty$. Scaling f by a positive value a (that is, the transformed function ab^t) does not affect the long-range behavior: whether the function tends to 0 or increases without bound depends solely on whether b is less than or greater than 1.

- The function $f(t) = b^t$ passes through $(0, 1)$, is always concave up, is either always increasing or always decreasing, and its range is the set of all positive real numbers. Among these properties, a vertical stretch by a positive value a only affects the y-intercept, which is instead $(0, a)$. If we include a vertical shift and write $g(t) = ab^t + c$, the biggest change is that the range of g is the set of all real numbers greater than c. In addition, the y-intercept of g is $(0, a + c)$.

In the situation where $a < 0$, several other changes are induced. Here, because $g(t) = ab^t$ is both a reflection of $f(t) = b^t$ across the x-axis and a vertical stretch by $|a|$, the function g is now always concave down. If $0 < b < 1$ so that f is always decreasing, then g (the reflected function) is now always increasing; if instead $b > 1$ so f is increasing, then g is decreasing. Finally, if $a < 0$, then the range of $g(t) = ab^t + c$ is the set of all real numbers c.

- An exponential function can be thought of as a function that changes at a rate proportional to itself, like how money grows with compound interest or the amount of a radioactive quantity decays. Newton's Law of Cooling says that the rate of change of an object's temperature is proportional to the *difference* between its own temperature and the temperature of its surroundings. This leads to the function that measures the difference between the object's temperature and room temperature being exponential, and hence the object's temperature itself is a vertically-shifted exponential function of the form $F(t) = ab^t + c$.

3.2.5 Exercises

1. If $b > 1$, what is the horizontal asymptote of $y = ab^t$ as $t \to -\infty$?

2. Find the long run behavior of each of the following functions.

    (a) As $x \to \infty$, $18(0.8)^x \to$ _____

    (b) As $t \to -\infty$, $9(2.2)^t \to$ _____

    (c) As $t \to \infty$, $0.6(2 + (0.1)^t) \to$ _____

3. Suppose t_0 is the t-coordinate of the point of intersection of the graphs below. Complete the statement below in order to correctly describe what happens to t_0 if the value of r (in the blue graph of $f(t) = a(1+r)^t$ below) is increased, and all other quantities remain the same.

As r increases, does the value of t_0 increase, decrease, or remain the same?

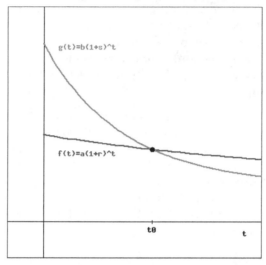

4. A can of soda has been in a refrigerator for several days; the refrigerator has temperature 41° Fahrenheit. Upon removal, the soda is placed on a kitchen table in a room with surrounding temperature 72°. Let $F(t)$ represent the soda's temperature in degrees Fahrenheit at time t in minutes, where $t = 0$ corresponds to the time the can is removed from the refrigerator. We know from Newton's Law of Cooling that F has form $F(t) = ab^t + c$ for some constants a, b, and c, where $0 < b < 1$.

    a. What is the numerical value of the soda's initial temperature? What is the value of $F(0)$ in terms of a, b, and c? What do these two observations tell us?

    b. What is the numerical value of the soda's long-term temperature? What is the long-term value of $F(t)$ in terms of a, b, and c? What do these two observations tell us?

    c. Using your work in (a) and (b), determine the numerical values of a and c.

    d. Suppose it can be determined that $b = 0.931$. What is the soda's temperature after 10 minutes?

5. Consider the graphs of the following four functions p, q, r, and s. Each is a shifted exponential function of the form $ab^t + c$.

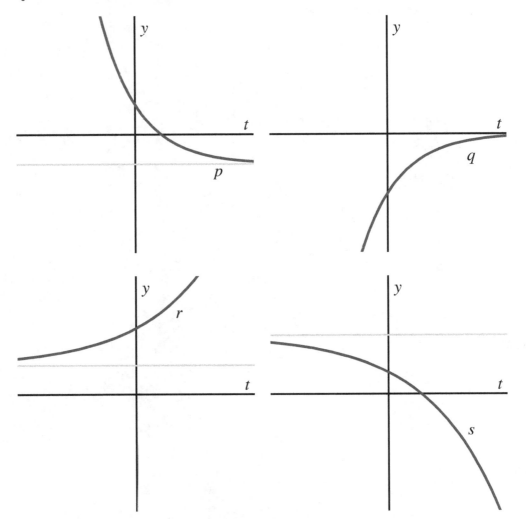

For each function p, q, r, and s, determine

- whether $a > 0$ or $a < 0$;
- whether $0 < b < 1$ or $b > 1$;
- whether $c > 0$, $c = 0$, or $c < 0$; and
- the range of the function in terms of c.

6. A cup of coffee has its temperature, $C(t)$, measured in degrees Celsius. When poured outdoors on a cold morning, its temperature is $C(0) = 95$. Ten minutes later, $C(10) = 80$. If the surrounding temperature outside is $0°$ Celsius, find a formula for a function $C(t)$ that models the coffee's temperature at time t.

    In addition, recall that we can convert between Celsius and Fahrenheit according to the equations $F = \frac{9}{5}C+32$ and $C = \frac{5}{9}(F-32)$. Use this information to also find a formula for $F(t)$, the coffee's Fahrenheit temperature at time t. What is similar and what is different regarding the functions $C(t)$ and $F(t)$?

3.3 The special number *e*

Motivating Questions

- Why can every exponential function of form $f(t) = b^t$ (where $b > 0$ and $b \neq 1$) be thought of as a horizontal scaling of a single special exponential function?

- What is the natural base *e* and what makes this number special?

We have observed that the behavior of functions of the form $f(t) = b^t$ is very consistent, where the only major differences depend on whether $b < 1$ or $b > 1$. Indeed, if we stipulate that $b > 1$, the graphs of functions with different bases b look nearly identical, as seen in the plots of p, q, r, and s in Figure 3.3.1.

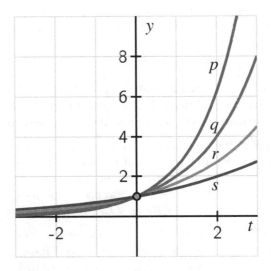

Figure 3.3.1: Plots of four different exponential functions of form b^t with $b > 1$.

Because the point $(0, 1)$ lies on the graph of each of the four functions in Figure 3.3.1, the functions cannot be vertical scalings of one another. However, it is possible that the functions are *horizontal* scalings of one another. This leads us to a natural question: might it be possible to find a single exponential function with a special base, say e, for which every other exponential function $f(t) = b^t$ can be expressed as a horizontal scaling of $E(t) = e^t$?

> **Preview Activity 3.3.1.** Open a new *Desmos* worksheet and define the following functions: $f(t) = 2^t$, $g(t) = 3^t$, $h(t) = (\frac{1}{3})^t$, and $p(t) = f(kt)$. After you define p, accept the slider for k, and set the range of the slider to be $-2 \leq k \leq 2$.
>
> a. By experimenting with the value of k, find a value of k so that the graph of $p(t) = f(kt) = 2^{kt}$ appears to align with the graph of $g(t) = 3^t$. What is the value of k?

b. Similarly, experiment to find a value of k so that the graph of $p(t) = f(kt) = 2^{kt}$ appears to align with the graph of $h(t) = (\frac{1}{3})^t$. What is the value of k?

c. For the value of k you determined in (a), compute 2^k. What do you observe?

d. For the value of k you determined in (b), compute 2^k. What do you observe?

e. Given any exponential function of the form b^t, do you think it's possible to find a value of k to that $p(t) = f(kt) = 2^{kt}$ is the same function as b^t? Why or why not?

3.3.1 The natural base e

In Preview Activity 3.3.1, we found that it appears possible to find a value of k so that given any base b, we can write the function b^t as the horizontal scaling of 2^t given by

$$b^t = 2^{kt}.$$

It's also apparent that there's nothing particularly special about "2": we could similarly write any function b^t as a horizontal scaling of 3^t or 4^t, albeit with a different scaling factor k for each. Thus, we might also ask: is there a *best* possible single base to use?

Through the central topic of the *rate of change* of a function, calculus helps us decide which base is best to use to represent all exponential functions. While we study *average* rate of change extensively in this course, in calculus there is more emphasis on the *instantaneous* rate of change. In that context, a natural question arises: is there a nonzero function that grows in such a way that its *height* is exactly how *fast* its height is increasing?

Amazingly, it turns out that the answer to this questions is "yes," and the function with this property is **the exponential function with the natural base**, denoted e^t. The number e (named in homage to the great Swiss mathematician Leonard Euler (1707-1783)) is complicated to define. Like π, e is an irrational number that cannot be represented exactly by a ratio of integers and whose decimal expansion never repeats. Advanced mathematics is needed in order to make the following formal definition of e.

Definition 3.3.2 The natural base, e. The number e is the infinite sum[1]

$$e = 1 + \frac{1}{1!} + \frac{1}{2!} + \frac{1}{3!} + \frac{1}{4!} + \cdots$$

From this, $e \approx 2.718281828$. ◊

For instance, $1 + \frac{1}{1} + \frac{1}{2} + \frac{1}{6} + \frac{1}{24} + \frac{1}{120} = \frac{163}{60} \approx 2.7167$ is an approximation of e generated by taking the first 6 terms in the infinite sum that defines it. Every computational device knows the number e and we will normally work with this number by using technology appropriately.

Initially, it's important to note that $2 < e < 3$, and thus we expect the function e^t to lie between 2^t and 3^t.

---

[1]Infinite sums are usually studied in second semester calculus.

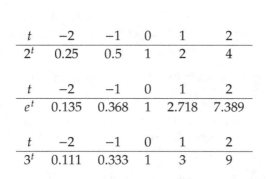

t	-2	-1	0	1	2
2^t	0.25	0.5	1	2	4

t	-2	-1	0	1	2
e^t	0.135	0.368	1	2.718	7.389

t	-2	-1	0	1	2
3^t	0.111	0.333	1	3	9

Table 3.3.3: Select outputs of 2^t, e^t, and 3^t reported to 3 decimal places.

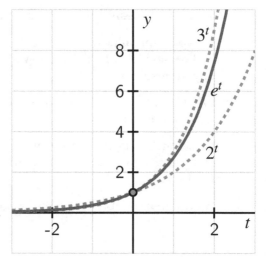

Figure 3.3.4: Plot of e^t along with 2^t and 3^t.

If we compare the graphs and some selected outputs of each function, as in Table 3.3.3 and Figure 3.3.4, we see that the function e^t satisfies the inequality

$$2^t < e^t < 3^t$$

for all positive values of t. When t is negative, we can view the values of each function as being reciprocals of powers of 2, e, and 3. For instance, since $2^2 < e^2 < 3^2$, it follows $\frac{1}{3^2} < \frac{1}{e^2} < \frac{1}{2^2}$, or

$$3^{-2} < e^{-2} < 2^{-2}.$$

Thus, for any $t < 0$,

$$3^t < e^t < 2^t$$

Like 2^t and 3^t, the function e^t passes through $(0, 1)$ is always increasing and always concave up, and its range is the set of all positive real numbers.

Activity 3.3.2. Recall from Section 1.3 that the average rate of change of a function f on an interval $[a, b]$ is

$$AV_{[a,b]} = \frac{f(b) - f(a)}{b - a}.$$

In Section 1.6, we also saw that if we instead think of the average rate of change of f on the interval $[a, a + h]$, the expression changes to

$$AV_{[a,a+h]} = \frac{f(a + h) - f(a)}{h}.$$

In this activity we explore the average rate of change of $f(t) = e^t$ near the points where $t = 1$ and $t = 2$.

In a new *Desmos* worksheet, let $f(t) = e^t$ and define the function A by the rule

$$A(h) = \frac{f(1 + h) - f(1)}{h}.$$

a. What is the meaning of $A(0.5)$ in terms of the function f and its graph?

b. Compute the value of $A(h)$ for at least 6 different small values of h, both positive and negative. For instance, one value to try might be $h = 0.0001$. Record a table of your results.

c. What do you notice about the values you found in (b)? How do they compare to an important number?

d. Explain why the following sentence makes sense: "The function e^t is increasing at an average rate that is about the same as its value on small intervals near $t = 1$."

e. Adjust your definition of A in *Desmos* by changing 1 to 2 so that

$$A(h) = \frac{f(2 + h) - f(2)}{h}.$$

How does the value of $A(h)$ compare to $f(2)$ for small values of h?

3.3.2 Why any exponential function can be written in terms of e

In Preview Activity 3.3.1, we saw graphical evidence that any exponential function $f(t) = b^t$ can be written as a horizontal scaling of the function $g(t) = 2^t$, plus we observed that there wasn't anything particularly special about 2^t. Because of the importance of e^t in calculus, we will choose instead to use the natural exponential function, $E(t) = e^t$ as the function we scale to generate any other exponential function $f(t) = b^t$. We claim that for any choice of $b > 0$ (with $b \neq 1$), there exists a horizontal scaling factor k such that $b^t = f(t) = E(kt) = e^{kt}$.

By the rules of exponents, we can rewrite this last equation equivalently as

$$b^t = (e^k)^t.$$

Since this equation has to hold for every value of t, it follows that $b = e^k$. Thus, our claim that we can scale $E(t)$ to get $f(t)$ requires us to show that regardless of the choice of the positive number b, there exists a single corresponding value of k such that $b = e^k$.

Given $b > 0$, we can always find a corresponding value of k such that $e^k = b$ because the function $f(t) = e^t$ passes the Horizontal Line Test, as seen in Figure 3.3.5. In Figure 3.3.5, we can think of b as a point on the positive vertical axis. From there, we draw a horizontal line over to the graph of $f(t) = e^t$, and then from the (unique) point of intersection we drop a vertical line to the x-axis. At that corresponding point on the x-axis we have found the input value k that corresponds to b. We see that there is always exactly one such k value that corresponds to each chosen b because $f(t) = e^t$ is always increasing, and any always increasing function passes the Horizontal Line Test.

It follows that the function $f(t) = e^t$ has an inverse function, and hence there must be some other function g such that writing $y = f(t)$ is the same as writing $t = g(y)$. This important function g will be developed in Section 3.4 and will enable us to find the value of k exactly

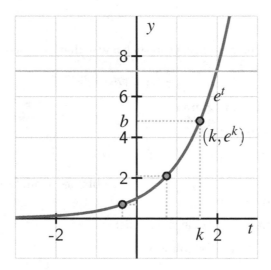

Figure 3.3.5: A plot of $f(t) = e^t$ along with several choices of positive constants b viewed on the vertical axis.

for a given b. For now, we are content to work with these observations graphically and to hence find estimates for the value of k.

Activity 3.3.3. By graphing $f(t) = e^t$ and appropriate horizontal lines, estimate the solution to each of the following equations. Note that in some parts, you may need to do some algebraic work in addition to using the graph.

a. $e^t = 2$

b. $e^{3t} = 5$

c. $2e^t - 4 = 7$

d. $3e^{0.25t} + 2 = 6$

e. $4 - 2e^{-0.7t} = 3$

f. $2e^{1.2t} = 1.5e^{1.6t}$

3.3.3 Summary

- Any exponential function $f(t) = b^t$ can be viewed as a horizontal scaling of $E(t) = e^t$ because there exists a unique constant k such that $E(kt) = e^{kt} = b^t = f(t)$ is true for every value of t. This holds since the exponential function e^t is always increasing, so given an output b there exists a unique input k such that $e^k = b$, from which it follows that $e^{kt} = b^t$.

- The natural base e is the special number that defines an increasing exponential function whose rate of change at any point is the same as its height at that point, a fact that is established using calculus. The number e turns out to be given exactly by an infinite sum and approximately by $e \approx 2.7182818$.

3.3.4 Exercises

1. Certain radioactive material decays in such a way that the mass remaining after t years is given by the function
$$m(t) = 175e^{-0.015t}$$
where $m(t)$ is measured in grams.

(a) Find the mass at time $t = 0$.

(b) How much of the mass remains after 15 years?

2. The graph of the function $f(x) = -e^x$ can be obtained from the graph of $g(x) = e^x$ by one of the following actions:

(a) reflecting the graph of $g(x)$ in the y-axis;

(b) reflecting the graph of $g(x)$ in the x-axis;

The range of the function $f(x)$ is $f(x) < A$, find A.

Is the domain of the function $f(x)$ still $(-\infty, \infty)$?

3. The graph of the function $f(x) = e^{-x} - 4$ can be obtained from the graph of $g(x) = e^x$ by two of the following actions:

(a) reflecting the graph of $g(x)$ in the y-axis;

(b) reflecting the graph of $g(x)$ in the x-axis;

(c) shifting the graph of $g(x)$ to the right 4 units;

(d) shifting the graph of $g(x)$ to the left 4 units;

(e) shifting the graph of $g(x)$ upward 4 units;

(f) shifting the graph of $g(x)$ downward 4 units;

(Please give your answer in the order the changes are applied, e.g. a first, then b second.)

The range of the function $f(x)$ is $f(x) > A$, find A.

Is the domain of the function $f(x)$ still $(-\infty, \infty)$?

4. Find the end behavior of the following function. As $t \to -\infty$, $11e^{0.14t} \to$ _____

5. Compute the following limit.
$$\lim_{t \to \infty} \left(12e^{-0.15t} + 17\right) = \underline{\hspace{3cm}}$$

6. When a single investment of principal, $\$P$, is invested in an account that returns interest at an annual rate of r (a decimal that corresponds to the percentage rate, such as $r = 0.05$ corresponding to 5%) that is compounded n times per year, the amount of money in the account after t years is given by $A(t) = P(1 + \frac{r}{n})^{nt}$.

Suppose we invest $\$P$ in an account that earns 8% annual interest. We investigate the effects of different rates of compounding.

   a. Compute $A(1)$ if interest is compounded quarterly $(n = 4)$.

   b. Compute $A(1)$ if interest is compounded monthly.

   c. Compute $A(1)$ if interest is compounded weekly.

   d. Compute $A(1)$ if interest is compounded daily.

   e. If we let the number of times that interest is compounded increase without bound, we say that the interest is "compounded continuously". When interest is compounded continuously, it turns out that the amount of money an account with initial investment \$$P$ after t years at an annual interest rate of r is $A(t) = Pe^{rt}$, where e is the natural base. Compute $A(1)$ in the same context as the preceding questions but where interest is compounded continuously.

   f. How much of a difference does continuously compounded interest make over interest compounded quarterly in one year's time? How does your answer change over 25 years' time?

7.   In *Desmos*, define the function $g(t) = e^{kt}$ and accept the slider for k. Set the range of the slider to $-2 \le k \le 2$, and assume that $k \ne 0$. Experiment with a wide range of values of k to see the effects of changing k.

   a. For what values of k is g always increasing? For what values of k is g always decreasing?

   b. For which value of k is the average rate of change of g on $[0, 1]$ greater: when $k = -0.1$ or when $k = -0.05$?

   c. What is the long-term behavior of g when $k < 0$? Why does this occur?

   d. Experiment with the slider to find a value of k for which $g(2) = \frac{1}{2}$. Test your estimate by computing e^{2k}. How accurate is your estimate?

8.   A can of soda is removed from a refrigerator at time $t = 0$ (in minutes) and its temperature, $F(t)$, in degrees Fahrenheit, is computed at regular intervals. Based on the data, a model is formulated for the object's temperature, given by

$$F(t) = 74.4 - 38.8e^{-0.05t}.$$

   a. What is the long-term behavior of the function $g(t) = e^{-0.05t}$? Why?

   b. What is the long-term behavior of the function $F(t) = 74.4 - 38.8e^{-0.05t}$? What is the meaning of this value in the physical context of the problem?

   c. What is the temperature of the refrigerator? Why?

   d. Compute the average rate of change of F on the intervals $[10, 20]$, $[20, 30]$, and $[30, 40]$. Write a careful sentence, with units, to explain the meaning of each, and write an additional sentence to describe any overall trends in how the average rate of change of F is changing.

3.4 What a logarithm is

Motivating Questions

- How is the base-10 logarithm defined?

- What is the "natural logarithm" and how is it different from the base-10 logarithm?

- How can we solve an equation that involves e to some unknown quantity?

In Section 1.7, we introduced the idea of an inverse function. The fundamental idea is that f has an inverse function if and only if there exists another function g such that f and g "undo" one another's respective processes. In other words, the process of the function f is reversible, and reversing f generates a related function g.

More formally, recall that a function $y = f(x)$ (where $f : A \rightarrow B$) has an inverse function if and only if there exists another function $g : B \rightarrow A$ such that $g(f(x)) = x$ for every x in A, and $f(g(y)) = y$ for every y in B. We know that given a function f, we can use the Horizontal Line Test to determine whether or not f has an inverse function. Finally, whenever a function f has an inverse function, we call its inverse function f^{-1} and know that the two equations $y = f(x)$ and $x = f^{-1}(y)$ say the same thing from different perspectives.

> **Preview Activity 3.4.1.** Let $P(t)$ be the "powers of 10" function, which is given by $P(t) = 10^t$.
>
> a. Complete Table 3.4.1 to generate certain values of P.
>
> | t | -3 | -2 | -1 | 0 | 1 | 2 | 3 |
> |---|---|---|---|---|---|---|---|
> | $y = P(t) = 10^t$ | | | | | | | |
>
> **Table 3.4.1:** Select values of the powers of 10 function.
>
> b. Why does P have an inverse function?
>
> c. Since P has an inverse function, we know there exists some other function, say L, such that writing "$y = P(t)$" says the exact same thing as writing "$t = L(y)$". In words, where P produces the result of raising 10 to a given power, the function L reverses this process and instead tells us the power to which we need to raise 10, given a desired result. Complete Table 3.4.2 to generate a collection of values of L.
>
> | y | 10^{-3} | 10^{-2} | 10^{-1} | 10^0 | 10^1 | 10^2 | 10^3 |
> |---|---|---|---|---|---|---|---|
> | $L(y)$ | | | | | | | |
>
> **Table 3.4.2:** Select values of the function L that is the inverse of P.

d. What are the domain and range of the function P? What are the domain and range of the function L?

3.4.1 The base-10 logarithm

The powers-of-10 function $P(t) = 10^t$ is an exponential function with base $b > 1$. As such, P is always increasing, and thus its graph passes the Horizontal Line Test, so P has an inverse function. We therefore know there exists some other function, L, such that writing $y = P(t)$ is equivalent to writing $t = L(y)$. For instance, we know that $P(2) = 100$ and $P(-3) = \frac{1}{1000}$, so it's equivalent to say that $L(100) = 2$ and $L(\frac{1}{1000}) = -3$. This new function L we call the *base* 10 *logarithm*, which is formally defined as follows.

Definition 3.4.3 Given a positive real number y, the **base-10 logarithm of** y is the power to which we raise 10 to get y. We use the notation "$\log_{10}(y)$" to denote the base-10 logarithm of y. ◊

The base-10 logarithm is therefore the inverse of the powers of 10 function. Whereas $P(t) = 10^t$ takes an input whose value is an exponent and produces the result of taking 10 to that power, the base-10 logarithm takes an input number we view as a power of 10 and produces the corresponding exponent such that 10 to that exponent is the input number.

In the notation of logarithms, we can now update our earlier observations with the functions P and L and see how exponential equations can be written in two equivalent ways. For instance,

$$10^2 = 100 \text{ and } \log_{10}(100) = 2 \tag{3.4.1}$$

each say the same thing from two different perspectives. The first says "100 is 10 to the power 2", while the second says "2 is the power to which we raise 10 to get 100". Similarly,

$$10^{-3} = \frac{1}{1000} \text{ and } \log_{10}\left(\frac{1}{1000}\right) = -3. \tag{3.4.2}$$

If we rearrange the statements of the facts in Equation (3.4.1), we can see yet another important relationship between the powers of 10 and base-10 logarithm function. Noting that $\log_{10}(100) = 2$ and $100 = 10^2$ are equivalent statements, and substituting the latter equation into the former, we see that

$$\log_{10}(10^2) = 2. \tag{3.4.3}$$

In words, Equation (3.4.3) says that "the power to which we raise 10 to get 10^2, is 2". That is, the base-10 logarithm function undoes the work of the powers of 10 function.

In a similar way, if we rearrange the statements in Equation (3.4.2), we can observe that by replacing -3 with $\log_{10}(\frac{1}{1000})$ we have

$$10^{\log_{10}(\frac{1}{1000})} = \frac{1}{1000}. \tag{3.4.4}$$

In words, Equation (3.4.4) says that "when 10 is raised to the power to which we raise 10 in order to get $\frac{1}{1000}$, we get $\frac{1}{1000}$".

We summarize the key relationships between the powers-of-10 function and its inverse, the base-10 logarithm function, more generally as follows.

$P(t) = 10^t$ **and** $L(y) = \log_{10}(y)$.

- The domain of P is the set of all real numbers and the range of P is the set of all positive real numbers.

- The domain of L is the set of all positive real numbers and the range of L is the set of all real numbers.

- For any real number t, $\log_{10}(10^t) = t$. That is, $L(P(t)) = t$.

- For any positive real number y, $10^{\log_{10}(y)} = y$. That is, $P(L(y)) = y$.

- $10^0 = 1$ and $\log_{10}(1) = 0$.

The base-10 logarithm function is like the sine or cosine function in this way: for certain special values, it's easy to know the value of the logarithm function. While for sine and cosine the familiar points come from specially placed points on the unit circle, for the base-10 logarithm function, the familiar points come from powers of 10. In addition, like sine and cosine, for all other input values, (a) calculus ultimately determines the value of the base-10 logarithm function at other values, and (b) we use computational technology in order to compute these values. For most computational devices, the command `log(y)` produces the result of the base-10 logarithm of y.

It's important to note that the logarithm function produces exact values. For instance, if we want to solve the equation $10^t = 5$, then it follows that $t = \log_{10}(5)$ is the exact solution to the equation. Like $\sqrt{2}$ or $\cos(1)$, $\log_{10}(5)$ is a number that is an exact value. A computational device can give us a decimal approximation, and we normally want to distinguish between the exact value and the approximate one. For the three different numbers here, $\sqrt{2} \approx 1.414$, $\cos(1) \approx 0.540$, and $\log_{10}(5) \approx 0.699$.

Activity 3.4.2. For each of the following equations, determine the exact value of the unknown variable. If the exact value involves a logarithm, use a computational device to also report an approximate value. For instance, if the exact value is $y = \log_{10}(2)$, you can also note that $y \approx 0.301$.

a. $10^t = 0.00001$

b. $\log_{10}(1000000) = t$

c. $10^t = 37$

d. $\log_{10}(y) = 1.375$

e. $10^t = 0.04$

f. $3 \cdot 10^t + 11 = 147$

g. $2\log_{10}(y) + 5 = 1$

3.4.2 The natural logarithm

The base-10 logarithm is a good starting point for understanding how logarithmic functions work because powers of 10 are easy to mentally compute. We could similarly consider the powers of 2 or powers of 3 function and develop a corresponding logarithm of base 2 or 3. But rather than have a whole collection of different logarithm functions, in the same way that we now use the function e^t and appropriate scaling to represent any exponential function, we develop a single logarithm function that we can use to represent any other logarithmic function through scaling. In correspondence with the natural exponential function, e^t, we now develop its inverse function, and call this inverse function the **natural logarithm**.

Definition 3.4.4 Given a positive real number y, the **natural logarithm of** y is the power to which we raise e to get y. We use the notation "$\ln(y)$" to denote the natural logarithm of y.

◊

We can think of the natural logarithm, $\ln(y)$, as the "base-e logarithm". For instance,

$$\ln(e^2) = 2$$

and

$$e^{\ln(-1)} = -1.$$

The former equation is true since "the power to which we raise e to get e^2 is 2"; the latter equation is true since "when we raise e to the power to which we raise e to get -1, we get -1". The key relationships between the natural exponential and the natural logarithm function are investigated in Activity 3.4.3.

Activity 3.4.3. Let $E(t) = e^t$ and $N(y) = \ln(y)$ be the natural exponential function and the natural logarithm function, respectively.

a. What are the domain and range of E?

b. What are the domain and range of N?

c. What can you say about $\ln(e^t)$ for every real number t?

d. What can you say about $e^{\ln(y)}$ for every positive real number y?

e. Complete Table 3.4.5 and Table 3.4.6 with both exact and approximate values of E and N. Then, plot the corresponding ordered pairs from each table on the axes provided in Figure 3.4.7 and connect the points in an intuitive way. When you plot the ordered pairs on the axes, in both cases view the first line of the table as generating values on the horizontal axis and the second line of the table as producing values on the vertical axis[1]; label each ordered pair you plot appropriately.

t	-2	-1	0	1	2
$E(t) = e^t$	$e^{-2} \approx 0.135$				

Table 3.4.5: Values of $y = E(t)$.

y	e^{-2}	e^{-1}	1	e^1	e^2
$N(y) = \ln(y)$	-2				

Table 3.4.6: Values of $t = N(y)$.

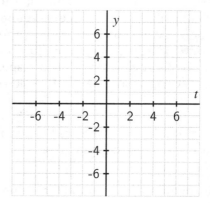

Figure 3.4.7: Axes for plotting data from Table 3.4.5 and Table 3.4.6 along with the graphs of the natural exponential and natural logarithm functions.

3.4.3 $f(t) = b^t$ revisited

In Section 3.1 and Section 3.2, we saw that that function $f(t) = b^t$ plays a key role in modeling exponential growth and decay, and that the value of b not only determines whether the function models growth ($b > 1$) or decay ($0 < b < 1$), but also how fast the growth or decay occurs. Furthermore, once we introduced the natural base e in Section 3.3, we realized that we could write every exponential function of form $f(t) = b^t$ as a horizontal scaling of the function $E(t) = e^t$ by writing

$$b^t = f(t) = E(kt) = e^{kt}$$

for some value k. Our development of the natural logarithm function in the current section enables us to now determine k exactly.

Example 3.4.8 Determine the exact value of k for which $f(t) = 3^t = e^{kt}$.

Solution. Since we want $3^t = e^{kt}$ to hold for every value of t and $e^{kt} = (e^k)^t$, we need to have $3^t = (e^k)^t$, and thus $3 = e^k$. Therefore, k is the power to which we raise e to get 3, which by definition means that $k = \ln(3)$. $\square$

In modeling important phenomena using exponential functions, we will frequently encounter equations where the variable is in the exponent, like in Example 3.4.8 where we had to solve $e^k = 3$. It is in this context where logarithms find one of their most powerful applications. Activity 3.4.4 provides some opportunities to practice solving equations involving the natural base, e, and the natural logarithm.

---

[1]Note that when we take this perspective for plotting the data in Table 3.4.6, we are viewing N as a function of t, writing $N(t) = \ln(t)$ in order to plot the function on the t-y axes

Activity 3.4.4. Solve each of the following equations for the exact value of the unknown variable. If there is no solution to the equation, explain why not.

a. $e^t = \frac{1}{10}$

b. $5e^t = 7$

c. $\ln(t) = -\frac{1}{3}$

d. $e^{1-3t} = 4$

e. $2\ln(t) + 1 = 4$

f. $4 - 3e^{2t} = 2$

g. $4 + 3e^{2t} = 2$

h. $\ln(5 - 6t) = -2$

3.4.4 Summary

- The base-10 logarithm of y, denoted $\log_{10}(y)$, is defined to be the power to which we raise 10 to get y. For instance, $\log_{10}(1000) = 3$, since $10^3 = 1000$. The function $L(y) = \log_{10}(y)$ is thus the inverse of the powers-of-10 function, $P(t) = 10^t$.

- The natural logarithm $N(y) = \ln(y)$ differs from the base-10 logarithm in that it is the logarithm with base e instead of 10, and thus $\ln(y)$ is the power to which we raise e to get y. The function $N(y) = \ln(y)$ is the inverse of the natural exponential function $E(t) = e^t$.

- The natural logarithm often enables us solve an equation that involves e to some unknown quantity. For instance, to solve $2e^{3t-4} + 5 = 13$, we can first solve for e^{3t-4} by subtracting 5 from each side and dividing by 2 to get

$$e^{3t-4} = 4.$$

This last equation says "e to some power is 4". We know that it is equivalent to say

$$\ln(4) = 3t - 4.$$

Since $\ln(4)$ is a number, we can solve this most recent linear equation for t. In particular, $3t = 4 + \ln(4)$, so

$$t = \frac{1}{3}(4 + \ln(4)).$$

3.4.5 Exercises

1. Express the equation in exponential form

   (a) $\ln 4 = x$ is equivalent to $e^A = B$. Find A and B.

   (b) $\ln x = 3$ is equivalent to $e^C = D$. Find C and D.

2. Evaluate the following expressions. Your answers must be exact and in simplest form.

   (a) $\ln e^9$

   (b) $e^{\ln 2}$

   (c) $e^{\ln \sqrt{3}}$

(d) $\ln(1/e^2)$

 3. Find the solution of the exponential equation

$$20e^x - 1 = 15$$

in terms of logarithms, or correct to four decimal places.

 4. Find the solution of the exponential equation

$$e^{2x+1} = 18$$

in terms of logarithms, or correct to four decimal places.

 5. Find the solution of the logarithmic equation

$$6 - \ln(5 - x) = 0$$

correct to four decimal places.

6. Recall that when a function $y = f(x)$ has an inverse function, the two equations $y = f(x)$ and $x = f^{-1}(y)$ say the same thing from different perspectives: the first equation expresses y in terms of x, while the second expresses x in terms of y. When $y = f(x) = e^x$, we know its inverse is $x = f^{-1}(y) = \ln(y)$. Through logarithms, we now have the ability to find the inverse of many different exponential functions. In particular, because exponential functions and their transformations are either always increasing or always decreasing, any function of the form $y = f(x) = ae^{-kx} + c$ will have an inverse function.

Find the inverse function for each given function by solving algebraically for x as a function of y. In addition, state the domain and range of the given function and the domain and range of of the inverse function.

a. $y = g(x) = e^{-0.25x}$ e. $y = u(x) = -5e^{3x-4} + 8$

b. $y = h(x) = 2e^x + 1$

       f. $y = w(x) = 3\ln(x) + 4$

c. $y = r(x) = 21 + 15e^{-0.1x}$

d. $y = s(x) = 72 - 40e^{-0.05x}$ g. $y = z(x) = -0.2\ln(2x - 5) + 1$

7. We've seen that any exponential function $f(t) = b^t$ ($b > 0$, $b \neq 1$) can be written in the form $f(t) = e^{kt}$ for some real number k, and this is because $f(t) = b^t$ is a horizontal scaling of the function $E(t) = e^t$. In this exercise, we explore how the natural logarithm can be scaled to achieve a logarithm of any base.

Let $b > 1$. Because the function $y = f(t) = b^t$ has an inverse function, it makes sense to define its inverse like we did when $b = 10$ or $b = e$. The **base-b logarithm**, denoted $\log_b(y)$ is defined to be the power to which we raise b to get y. Thus, writing $y = f(t) = b^t$ is the same as writing $t = f^{-1}(y) = \log_b(y)$.

In *Desmos*, the natural logarithm function is given by $\ln(t)$, while the base-10 logarithm by $\log(t)$. To get a logarithm of a different base, such as a base-2 logarithm, type $\log_2(t)$ (the underscore will generate a subscript; then use the right arrow to get out of subscript mode).

In a new *Desmos* worksheet, enter V(t) = k * ln(t) and accept the slider for k. Set the lower and upper bounds for the slider to 0.01 and 15, respectively.

    a. Define $f(t) = \log_2(t)$ in *Desmos*. Can you find a value of k for which $\log_2(t) = k \ln(t)$? If yes, what is the value? If not, why not?

    b. Repeat (a) for the functions $g(t) = \log_3(t)$, $h(t) = \log_5(t)$, and $p(t) = \log_{1.25}(t)$. What pattern(s) do you observe?

    c. True or false: for any value of $b > 1$, the function $\log_b(t)$ can be viewed as a vertical scaling of $\ln(t)$.

    d. Compute the following values: $\frac{1}{\ln(2)}$, $\frac{1}{\ln(3)}$, $\frac{1}{\ln(5)}$, and $\frac{1}{\ln(1.25)}$. What do you notice about these values compared to those of k you found in (a) and (b)?

8.   A can of soda is removed from a refrigerator at time $t = 0$ (in minutes) and its temperature, $F(t)$, in degrees Fahrenheit, is computed at regular intervals. Based on the data, a model is formulated for the object's temperature, given by

$$F(t) = 74.4 - 38.8e^{-0.05t}.$$

    a. Determine the exact time when the soda's temperature is 50°.

    b. Is there ever a time when the soda's temperature is 36°? Why or why not?

    c. For the model, its domain is the set of all positive real numbers, $t > 0$. What is its range?

    d. Find a formula for the inverse of the function $y = F(t)$. What is the meaning of this function?

3.5 Properties and applications of logarithmic functions

Motivating Questions

- What structural rules do logarithms obey that are similar to rules for exponents?

- What are the key properties of the graph of the natural logarithm function?

- How do logarithms enable us to solve exponential equations?

Logarithms arise as inverses of exponential functions. In addition, we have motivated their development by our desire to solve exponential equations such as $e^k = 3$ for k. Because of the inverse relationship between exponential and logarithmic functions, there are several important properties logarithms have that are analogous to ones held by exponential functions. We will work to develop these properties and then show how they are useful in applied settings.

Preview Activity 3.5.1. In the following questions, we investigate how $\log_{10}(a \cdot b)$ can be equivalently written in terms of $\log_{10}(a)$ and $\log_{10}(b)$.

a. Write $10^x \cdot 10^y$ as 10 raised to a single power. That is, complete the equation

$$10^x \cdot 10^y = 10^\square$$

by filling in the box with an appropriate expression involving x and y.

b. What is the simplest possible way to write $\log_{10} 10^x$? What about the simplest equivalent expression for $\log_{10} 10^y$?

c. Explain why each of the following three equal signs is valid in the sequence of equalities:

$$\log_{10}(10^x \cdot 10^y) = \log_{10}(10^{x+y})$$
$$= x + y$$
$$= \log_{10}(10^x) + \log_{10}(10^y).$$

d. Suppose that a and b are positive real numbers, so we can think of a as 10^x for some real number x and b as 10^y for some real number y. That is, say that $a = 10^x$ and $b = 10^y$. What does our work in (c) tell us about $\log_{10}(ab)$?

3.5.1 Key properties of logarithms

In Preview Activity 3.5.1, we considered an argument for why $\log_{10}(ab) = \log_{10}(a) + \log_{10}(b)$ for any choice of positive numbers a and b. In what follows, we develop this and other properties of the natural logarithm function; similar reasoning shows the same properties hold for logarithms of any base.

Let a and b be any positive real numbers so that $x = \ln(a)$ and $y = \ln(b)$ are both defined. Observe that we can rewrite these two equations using the definition of the natural logarithm so that

$$a = e^x \text{ and } b = e^y.$$

Using substitution, we can now say that

$$\ln(a \cdot b) = \ln(e^x \cdot e^y).$$

By exponent rules, we know that $\ln(e^x \cdot e^y) = \ln(e^{x+y})$, and because the natural logarithm and natural exponential function are inverses, $\ln(e^{x+y}) = x + y$. Combining the three most recent equations,

$$\ln(a \cdot b) = x + y.$$

Finally, recalling that $x = \ln(a)$ and $y = \ln(b)$, we have shown that

$$\ln(a \cdot b) = \ln(a) + \ln(b)$$

for any choice of positive real numbers a and b.

A similar property holds for $\ln(\frac{a}{b})$. By nearly the same argument, we can say that

$$\ln\left(\frac{a}{b}\right) = \ln\left(\frac{e^x}{e^y}\right)$$
$$= \ln\left(e^{x-y}\right)$$
$$= x - y$$
$$= \ln(a) - \ln(b).$$

We have thus shown the following general principles.

> **Logarithms of products and quotients.**
>
> For any positive real numbers a and b,
> - $\ln(a \cdot b) = \ln(a) + \ln(b)$
> - $\ln\left(\frac{a}{b}\right) = \ln(a) - \ln(b)$

Because positive integer exponents are a shorthand way to express repeated multiplication, we can use the multiplication rule for logarithms to think about exponents as well. For example,

$$\ln(a^3) = \ln(a \cdot a \cdot a),$$

and by repeated application of the rule for the natural logarithm of a product, we see

$$\ln(a^3) = \ln(a) + \ln(a) + \ln(a) = 3\ln(a).$$

A similar argument works to show that for every natural number n,

$$\ln(a^n) = n\ln(a).$$

More sophisticated mathematics can be used to prove that the following property holds for every real number exponent t.

> **Logarithms of exponential expressions.**
>
> For any positive real number a and any real number t,
>
> $$\ln(a^t) = t\ln(a).$$

The rule that $\ln(a^t) = t\ln(a)$ is extremely powerful: by working with logarithms appropriately, it enables us to move from having a variable in an exponential expression to the variable being part of a linear expression. Moreover, it enables us to solve exponential equations exactly, regardless of the base involved.

Example 3.5.1 Solve the equation $7 \cdot 3^t - 1 = 5$ exactly for t.

Solution.   To solve for t, we first solve for 3^t. Adding 1 to both sides and dividing by 7, we find that $3^t = \frac{6}{7}$. Next, we take the natural logarithm of both sides of the equation. Doing so, we have

$$\ln\left(3^t\right) = \ln\left(\frac{6}{7}\right).$$

Applying the rule for the logarithm of an exponential expression on the left, we see that $t\ln(3) = \ln\left(\frac{6}{7}\right)$. Both $\ln(3)$ and $\ln\left(\frac{6}{7}\right)$ are simply numbers, and thus we conclude that

$$t = \frac{\ln(3)}{\ln\left(\frac{6}{7}\right)}.$$

$\square$

The approach used in Example 3.5.1 works in a wide range of settings: any time we have an exponential equation of the form $p \cdot q^t + r = s$, we can solve for t by first isolating the exponential expression q^t and then by taking the natural logarithm of both sides of the equation.

> **Activity 3.5.2.** Solve each of the following equations exactly and then find an estimate that is accurate to 5 decimal places.
>
> a. $3^t = 5$
>
> b. $4 \cdot 2^t - 2 = 3$
>
> c. $3.7 \cdot (0.9)^{0.3t} + 1.5 = 2.1$
>
> d. $72 - 30(0.7)^{0.05t} = 60$
>
> e. $\ln(t) = -2$
>
> f. $3 + 2\log_{10}(t) = 3.5$

3.5.2 The graph of the natural logarithm

As the inverse of the natural exponential function $E(x) = e^x$, we have already established that the natural logarithm $N(x) = \ln(x)$ has the set of all positive real numbers as its domain and the set of all real numbers as its range. In addition, being the inverse of $E(x) = e^x$, we know that when we plot the natural logarithm and natural exponential functions on the same coordinate axes, their graphs are reflections of one another across the line $y = x$, as seen in Figure 3.5.2 and Figure 3.5.3.

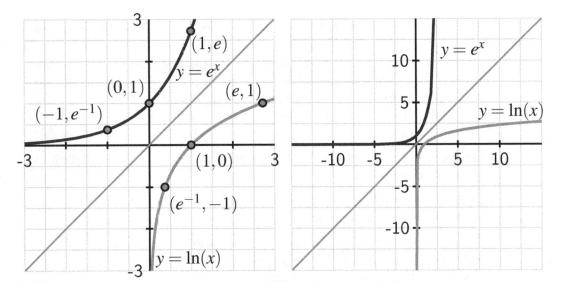

Figure 3.5.2: The natural exponential and natural logarithm functions on the interval $[-3,3]$.

Figure 3.5.3: The natural exponential and natural logarithm functions on the interval $[-15,15]$.

Indeed, for any point (a,b) that lies on the graph of $E(x) = e^x$, it follows that the point (b,a) lies on the graph of the inverse $N(x) = \ln(x)$. From this, we see several important properties of the graph of the logarithm function.

The graph of $y = \ln(x)$.

The graph of $y = \ln(x)$

- passes through the point $(1,0)$;

- is always increasing;

- is always concave down; and

- increases without bound.

Because the graph of $E(x) = e^x$ increases more and more rapidly as x increases, the graph of $N(x) = \ln(x)$ increases more and more slowly as x increases. Even though the natural logarithm function grows very slowly, it does grow without bound because we can make $\ln(x)$ as large as we want by making x sufficiently large. For instance, if we want x such that $\ln(x) = 100$, we choose $x = e^{100}$, since $\ln(e^{100}) = 100$.

While the natural exponential function and the natural logarithm (and transformations of these functions) are connected and have certain similar properties, it's also important to be able to distinguish between behavior that is fundamentally exponential and fundamentally logarithmic.

Activity 3.5.3. In the questions that follow, we compare and contrast the properties and behaviors of exponential and logarithmic functions.

a. Let $f(t) = 1 - e^{-(t-1)}$ and $g(t) = \ln(t)$. Plot each function on the same set of coordinate axes. What properties do the two functions have in common? For what properties do the two functions differ? Consider each function's domain, range, t-intercept, y-intercept, increasing/decreasing behavior, concavity, and long-term behavior.

b. Let $h(t) = a - be^{-k(t-c)}$, where $a, b, c,$ and k are positive constants. Describe h as a transformation of the function $E(t) = e^t$.

c. Let $r(t) = a + b\ln(t - c)$, where $a, b,$ and c are positive constants. Describe r as a transformation of the function $L(t) = \ln(t)$.

d. Data for the height of a tree is given in the Table 3.5.4; time t is measured in years and height is given in feet. At http://gvsu.edu/s/0yy, you can find a *Desmos* worksheet with this data already input.

t	1	2	3	4	5	6	7	8	9	10	11
$h(t)$	6	9.5	13	15	16.5	17.5	18.5	19	19.5	19.7	19.8

Table 3.5.4: The height of a tree as a function of time t in years.

Do you think this data is better modeled by a logarithmic function of form $p(t) = a + b\ln(t - c)$ or by an exponential function of form $q(t) = m + ne^{-rt}$. Provide reasons based in how the data appears and how you think a tree grows, as well as by experimenting with sliders appropriately in *Desmos*. (Note: you may need to adjust the upper and lower bounds of several of the sliders in order to match the data well.)

3.5.3 Putting logarithms to work

We've seen in several different settings that the function e^{kt} plays a key role in modeling phenomena in the world around us. We also understand that the value of k controls whether e^{kt} is increasing ($k > 0$) or decreasing ($k < 0$) and how fast the function is increasing or decreasing. As such, we often need to determine the value of k from data that is presented to us; doing so almost always requires the use of logarithms.

Example 3.5.5 A population of bacteria cells is growing at a rate proportionate to the number of cells present at a given time t (in hours). Suppose that the number of cells, P, in the population is measured in millions of cells and we know that $P(0) = 2.475$ and $P(10) = 4.298$. Find a model of the form $P(t) = Ae^{kt}$ that fits this data and use it to determine the value of k and how long it will take for the population to reach 1 billion cells.

Solution. Since the model has form $P(t) = Ae^{kt}$, we know that $P(0) = A$. Because we are given that $P(0) = 2.475$, this shows that $A = 2.475$. To find k, we use the fact that

$P(10) = 4.298$. Applying this information, $A = 2.475$, and the form of the model, $P(t) = Ae^{kt}$, we see that

$$4.298 = 2.475e^{k \cdot 10}.$$

To solve for k, we first isolate e^{10k} by dividing both sides by 2.475 to get

$$e^{10k} = \frac{4.298}{2.475}.$$

Taking the natural logarithm of each side, we find

$$10k = \ln\left(\frac{4.298}{2.475}\right),$$

and thus $k = \frac{1}{10}\ln\left(\frac{4.298}{2.475}\right) \approx 0.05519$.

To determine how long it takes for the population to reach 1 billion cells, we need to solve the equation $P(t) = 1000$. Using our preceding work to find A and k, we know that we need to solve the equation

$$1000 = 2.475e^{\frac{1}{10}\ln\left(\frac{4.298}{2.475}\right)t}.$$

We divide both sides by 2.475 to get $e^{\frac{1}{10}\ln\left(\frac{4.298}{2.475}\right)t} = \frac{1000}{2.475}$, and after taking the natural logarithm of each side, we see

$$\frac{1}{10}\ln\left(\frac{4.298}{2.475}\right)t = \ln\left(\frac{1000}{2.475}\right),$$

so that

$$t = \frac{10\ln\left(\frac{1000}{2.475}\right)}{\ln\left(\frac{4.298}{2.475}\right)} \approx 108.741.$$

□

Activity 3.5.4. Solve each of the following equations for the exact value of k.

a. $41 = 50e^{-k \cdot 7}$

b. $65 = 34 + 47e^{-k \cdot 45}$

c. $7e^{2k-1} + 4 = 32$

d. $\frac{5}{1+2e^{-10k}} = 4$

3.5.4 Summary

- There are three fundamental rules for exponents given nonzero base a and exponents m and n:

$$a^m \cdot a^n = a^{m+n}, \frac{a^m}{a^n} = a^{m-n}, \text{ and } (a^m)^n = a^{mn}.$$

For logarithms[1], we have the following analogous structural rules for positive real numbers a and b and any real number t:

$$\ln(a \cdot b) = \ln(a) + \ln(b), \ln\left(\frac{a}{b}\right) = \ln(a) - \ln(b), \text{ and } \ln(a^t) = t\ln(a).$$

---

[1]We state these rules for the natural logarithm, but they hold for any logarithm of any base.

- The natural logarithm's domain is the set of all positive real numbers and its range is the set of all real numbers. Its graph passes through $(1, 0)$, is always increasing, is always concave down, and increases without bound.

- Logarithms are very important in determining values that arise in equations of the form

$$a^b = c,$$

where a and c are known, but b is not. In this context, we can take the natural logarithm of both sides of the equation to find that

$$\ln(a^b) = \ln(c)$$

and thus $b \ln(a) = \ln(c)$, so that $b = \frac{\ln(c)}{\ln(a)}$.

3.5.5 Exercises

1. Solve for x: $3^x = 38$

2. Solve for x: $6 \cdot 4^{4x-4} = 65$

3. Find the solution of the exponential equation $11 + 5^{5x} = 16$ correct to at least four decimal places.

4. Find the solution of the exponential equation $1000(1.04)^{2t} = 50000$ in terms of logarithms, or correct to four decimal places.

5. Find the solution of the logarithmic equation $19 - \ln(3 - x) = 0$ correct to four decimal places.

6. For a population that is growing exponentially according to a model of the form $P(t) = Ae^{kt}$, the **doubling time** is the amount of time that it takes the population to double. For each population described below, assume the function is growing exponentially according to a model $P(t) = Ae^{kt}$, where t is measured in years.

   a. Suppose that a certain population initially has 100 members and doubles after 3 years. What are the values of A and k in the model?

   b. A different population is observed to satisfy $P(4) = 250$ and $P(11) = 500$. What is the population's doubling time? When will 2000 members of the population be present?

   c. Another population is observed to have doubling time $t = 21$. What is the value of k in the model?

   d. How is k related to a population's doubling time, regardless of how long the doubling time is?

7. A new car is purchased for $28000. Exactly 1 year later, the value of the car is $23200. Assume that the car's value in dollars, V, t years after purchase decays exponentially according to a model of form $V(t) = Ae^{-kt}$.

   a. Determine the exact values of A and k in the model.

   b. How many years will it take until the car's value is $10000?

    c. Suppose that rather than having the car's value decay all the way to $0, the lowest dollar amount its value ever approaches is $500. Explain why a model of the form $V(t) = Ae^{-kt} + c$ is more appropriate.

    d. Under the original assumptions ($V(0) = 28000$ and $V(1) = 23200$) along with the condition in (c) that the car's value will approach $500 in the long-term, determine the exact values of A, k, and c in the model $V(t) = Ae^{-kt} + c$. Are the values of A and k the same or different from the model explored in (a)? Why?

8.   In Exercise 3.4.5.7, we explored graphically how the function $y = \log_b(x)$ can be thought of as a vertical stretch of the nautral logarithm, $y = \ln(x)$. In this exercise, we determine the exact value of the vertical stretch that is needed.

Recall that $\log_b(x)$ is the power to which we raise b to get x.

    (a) Write the equation $y = \log_b(x)$ as an equivalent equation involving exponents with no logarithms present.

    (b) Take the equation you found in (a) and take the natural logarithm of each side.

    (c) Use rules and properties of logarithms appropriately to solve the equation from (b) for y. Your result here should express y in terms of $\ln(x)$ and $\ln(b)$.

    (d) Recall that $y = \log_b(x)$. Explain why the following equation (often called the Golden Rule for Logarithms) is true:

$$\log_b(x) = \frac{\ln(x)}{\ln(b)}.$$

    (e) What is the value of k that allows us to express the function $y = \log_b(x)$ as a vertical stretch of the function $y = \ln(x)$?

3.6 Modeling temperature and population

Motivating Questions

- What roles do the parameters a, k, and c play in how the function $F(t) = c + ae^{-kt}$ models the temperature of an object that is cooling or warming in its surroundings?

- How can we use an exponential function to more realistically model a population whose growth levels off?

We've seen that exponential functions can be used to model several different important phenomena, such as the growth of money due to continuously compounded interest, the decay of radioactive quanitities, and the temperature of an object that is cooling or warming due to its surroundings. From initial work with functions of the form $f(t) = ab^t$ where $b > 0$ and $b \neq 1$, we found that shifted exponential functions of form $g(t) = ab^t + c$ are also important. Moreover, the special base e allows us to represent all of these functions through horizontal scaling by writing

$$g(t) = ae^{kt} + c \tag{3.6.1}$$

where k is the constant such that $e^k = b$. Functions of the form of Equation (3.6.1) are either always increasing or always decreasing, always have the same concavity, are defined on the set of all real numbers, and have their range as the set of all real numbers greater than c or all real numbers less than c. In whatever setting we are using a model of this form, the crucial task is to identify the values of a, k, and c; that endeavor is the focus of this section.

We have also begun to see the important role that logarithms play in work with exponential models. The natural logarithm is the inverse of the natural exponential function and satisfies the important rule that $\ln(b^k) = k\ln(b)$. This rule enables us to solve equations with the structure $a^k = b$ for k in the context where a and b are known but k is not. Indeed, we can first take the natural log of both sides of the equation to get

$$\ln(a^k) = \ln(b),$$

from which it follows that $k\ln(a) = \ln(b)$, and therefore

$$k = \frac{\ln(b)}{\ln(a)}.$$

Finding k is often central to determining an exponential model, and logarithms make finding the exact value of k possible.

In Preview Activity 3.6.1, we revisit some key algebraic ideas with exponential and logarithmic equations in preparation for using these concepts in models for temperature and population.

> **Preview Activity 3.6.1.** In each of the following situations, determine the exact value of the unknown quantity that is identified.
>
> a. The temperature of a warming object in an oven is given by $F(t) = 275 - 203e^{-kt}$,

and we know that the object's temperature after 20 minutes is $F(20) = 101$. Determine the exact value of k.

b. The temperature of a cooling object in a refrigerator is modeled by $F(t) = a + 37.4e^{-0.05t}$, and the temperature of the refrigerator is 39.8°. By thinking about the long-term behavior of $e^{-0.05t}$ and the long-term behavior of the object's temperature, determine the exact value of a.

c. Later in this section, we'll learn that one model for how a population grows over time can be given by a function of the form

$$P(t) = \frac{A}{1 + Me^{-kt}}.$$

Models of this form lead naturally to equations that have structure like

$$3 = \frac{10}{1 + x}. \tag{3.6.2}$$

Solve Equation (3.6.2) for the exact value of x.

d. Suppose that $y = a + be^{-kt}$. Solve for t in terms of a, b, k, and y. What does this new equation represent?

3.6.1 Newton's Law of Cooling revisited

In Section 3.2, we learned that Newton's Law of Cooling, which states that an object's temperature changes at a rate proportional to the difference between its own temperature and the surrounding temperature, results in the object's temperature being modeled by functions of the form $F(t) = ab^t + c$. In light of our subsequent work in Section 3.3 with the natural base e, as well as the fact that $0 < b < 1$ in this model, we know that Newton's Law of Cooling implies that the object's temperature is modeled by a function of the form

$$F(t) = ae^{-kt} + c \tag{3.6.3}$$

for some constants a, c, and k, where $k > 0$.

From Equation (3.6.3), we can determine several different characteristics of how the constants a, b, and k are connected to the behavior of F by thinking about what happens at $t = 0$, at one additional value of t, and as t increases without bound. In particular, note that e^{-kt} will tend to 0 as t increases without bound.

Modeling temperature with Newton's Law of Cooling.

For the function $F(t) = ae^{-kt} + c$ that models the temperature of a cooling or warming object, the constants a, c, and k play the following roles.

- Since e^{-kt} tends to 0 as t increases without bound, $F(t)$ tends to c as t increases without bound, and thus c represents the temperature of the object's surroundings.

- Since $e^0 = 1$, $F(0) = a + c$, and thus the object's initial temperature is $a + c$. Said differently, a is the difference between the object's initial temperature and the temperature of the surroundings.

- Once we know the values of a and c, the value of k is determined by knowing the value of the temperature function $F(t)$ at one nonzero value of t.

Activity 3.6.2. A can of soda is initially at room temperature, $72.3°$ Fahrenheit, and at time $t = 0$ is placed in a refrigerator set at $37.7°$. In addition, we know that after 30 minutes, the soda's temperature has dropped to $59.5°$.

a. Use algebraic reasoning and your understanding of the physical situation to determine the exact values of a, c, and k in the model $F(t) = ae^{-kt} + c$. Write at least one careful sentence to explain your thinking.

b. Determine the exact time the object's temperature is $42.4°$. Clearly show your algebraic work and thinking.

c. Find the average rate of change of F on the interval $[25, 30]$. What is the meaning (with units) of this value?

d. If everything stayed the same except the value of $F(0)$, and instead $F(0) = 65$, would the value of k be larger or smaller? Why?

3.6.2 A more realistic model for population growth

If we assume that a population grows at a rate that is proportionate to the size of the population, it follows that the population grows exponentially according to the model

$$P(t) = Ae^{kt}$$

where A is the initial population and k is tied to the rate at which the population grows. Since $k > 0$, we know that e^{kt} is an always increasing, always concave up function that grows without bound. While $P(t) = Ae^{kt}$ may be a reasonable model for how a population grows when it is relatively small, because the function grows without bound as time increases, it can't be a realistic long-term representation of what happens in reality. Indeed, whether it is the number of fish who can survive in a lake, the number of cells in a petri dish, or the number of human beings on earth, the size of the surroundings and the limitations of resources will keep the population from being able to grow without bound.

In light of these observations, a different model is needed for population, one that grows exponentially at first, but that levels off later. Calculus can be used to develop such a model, and the resulting function is usually called the **logistic function**, which has form

$$P(t) = \frac{A}{1 + Me^{-kt}}, \tag{3.6.4}$$

where A, M, and k are positive constants. Since $k > 0$, it follows that $e^{-kt} \to 0$ as t increases without bound, and thus the denominator of P approaches 1 as time goes on. Thus, we observe that $P(t)$ tends to A as t increases without bound. We sometimes refer to A as the *carrying capacity* of the population.

Activity 3.6.3. In *Desmos*, define $P(t) = \frac{A}{1+Me^{-kt}}$ and accept sliders for A, M, and k. Set the slider ranges for these parameters as follows: $0.01 \le A \le 10$; $0.01 \le M \le 10$; $0.01 \le k \le 5$.

    a. Sketch a typical graph of $P(t)$ on the axes provided and write several sentences to explain the effects of A, M, and k on the graph of P.

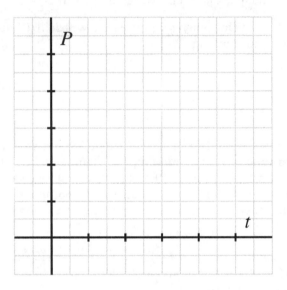

Figure 3.6.1: Axes for plotting a typical logistic function P.

    b. On a typical logistic graph, where does it appear that the population is growing most rapidly? How is this value connected to the carrying capacity, A?

    c. How does the function $1 + Me^{-kt}$ behave as t decreases without bound? What is the algebraic reason that this occurs?

    d. Use your *Desmos* worksheet to find a logistic function P that has the following properties: $P(0) = 2$, $P(2) = 4$, and $P(t)$ approaches 9 as t increases without bound. What are the approximate values of A, M, and k that make the function P fit these criteria?

Activity 3.6.4. Suppose that a population of animals that lives on an island (measured in thousands) is known to grow according to the logistic model, where t is measured in years. We know the following information: $P(0) = 2.45$, $P(3) = 4.52$, and as t increases without bound, $P(t)$ approaches 11.7.

    a. Determine the exact values of A, M, and k in the logistic model

$$P(t) = \frac{A}{1 + Me^{-kt}}.$$

    Clearly show your algebraic work and thinking.

    b. Plot your model from (a) and check that its values match the desired characteristics. Then, compute the average rate of change of P on the intervals $[0, 2]$, $[2, 4]$, $[4, 6]$, and $[6, 8]$. What is the meaning (with units) of the values you've found? How is the population growing on these intervals?

    c. Find the exact time value when the population will be 10 (thousand). Show your algebraic work and thinking.

3.6.3 Summary

- When a function of form $F(t) = c + ae^{-kt}$ models the temperature of an object that is cooling or warming in its surroundings, the temperature of the surroundings is c because $e^{-kt} \to 0$ as time goes on, the object's initial temperature is $a + c$, and the constant k is connected to how rapidly the object's temperature changes. Once a and c are known, the constant k can be determined by knowing the temperature at one additional time, t.

- Because the exponential function $P(t) = Ae^{kt}$ grows without bound as t increases, such a function is not a realistic model of a population that we expect to level off as time goes on. The logistic function

$$P(t) = \frac{A}{1 + Me^{-kt}}$$

more appropriately models a population that grows roughly exponentially when P is small but whose size levels off as it approaches the carrying capacity of the surrounding environment, which is the value of the constant A.

3.6.4 Exercises

1.   Newton's law of cooling states that the temperature of an object changes at a rate proportional to the difference between its temperature and that of its surroundings. Suppose that the temperature of a cup of coffee obeys Newton's law of cooling. If the coffee has a temperature of 210 degrees Fahrenheit when freshly poured, and 1.5 minutes later has cooled to 195 degrees in a room at 78 degrees, determine when the coffee reaches a temperature of 155 degrees.

2. The total number of people infected with a virus often grows like a logistic curve. Sup- pose that 25 people originally have the virus, and that in the early stages of the virus (with time, t, measured in weeks), the number of people infected is increasing exponentially with $k = 1.8$. It is estimated that, in the long run, approximately 7250 people become infected.

(a) Use this information to find a logistic function to model this situation.

(b) Sketch a graph of your answer to part (a). Use your graph to estimate the length of time until the rate at which people are becoming infected starts to decrease. What is the vertical coordinate at this point?

3. The town of Sickville, with a population of 9310 is exposed to the Blue Moon Virus, against which there is no immunity. The number of people infected when the virus is detected is 30. Suppose the number of infections grows logistically, with $k = 0.18$.

Find A.

Find the formula for the number of people infected after t days.

Find the number of people infected after 30 days.

4. A glass filled with ice and water is set on a table in a climate-controlled room with constant temperature of 71° Fahrenheit. A temperature probe is placed in the glass, and we find that the following temperatures are recorded (at time t in minutes).

t	0	20
$F(t)$	34.2	41.7

a. Make a rough sketch of how you think the temperature graph should appear. Is the temperature function always increasing? always decreasing? always concave up? always concave down? what's its long-range behavior?

b. By desribing F as a transformation of e^t, explain why a function of form $F(t) = c - ae^{-kt}$, where a, c, and k are positive constants is an appropriate model for how we expect the temperature function to behave.

c. Use the given information to determine the exact values of a, c, and k in the model $F(t) = c - ae^{-kt}$.

d. Determine the exact time when the water's temperature is 60°.

5. A popular cruise ship sets sail in the Gulf of Mexico with 5000 passengers and crew on board. Unfortunately, a five family members who board the ship are carrying a highly contagious virus. After interacting with many other passengers in the first few hours of the cruise, all five of them get very sick.

Let $S(t)$ be the number of people who have acquired the virus t days after the ship has left port. It turns out that a logistic function is a good model for S, and thus we assume that

$$S(t) = \frac{A}{1 + Me^{-kt}}$$

for some positive constants A, M, and k. Suppose that after 1 day, 20 people have gotten the virus.

a. Recall we know that $S(0) = 5$ and $S(1) = 20$. In addition, assume that 5000 is the number of people who will eventually get sick. Use this information determine the exact values of A, M, and k in the logistic model.

b. How many days will it take for 4000 of the people on the cruise ship to have acquired the virus?

c. Compute the average rate of change of S on the intervals $[1,2]$, $[3,4]$, and $[5,7]$. What is the meaning of each of these values (with units) in the context of the question, and what trend(s) do you observe in these average rates of change?

6. A closed tank with an inflow and outflow contains a 100 liters of saltwater solution. Let the amount of salt in the tank at time t (in minutes) be given by the function $A(t)$, whose output is measured in grams. At time $t = 0$ there is an initial amount of salt present in the tank, and the inflow line also carries a saltwater mixture to the tank at a fixed rate; the outflow occurs at the same rate and carries a perfectly mixed solution out of the tank. Because of these conditions, the volume of solution in the tank stays fixed over time, but the amount of salt possibly changes.

It turns out that the problem of determining the amount of salt in the tank at time t is similar to the problem of determining the temperature of a warming or cooling object, and that the function $A(t)$ has form

$$A(t) = ae^{-kt} + c$$

for constants a, c, and k. Suppose that for a particular set of conditions, we know that

$$A(t) = -500e^{-0.25t} + 750.$$

Again, $A(t)$ measures the amount of salt in the tank after t minutes.

a. How much salt is in the tank initially?

b. In the long run, how much salt do we expect to eventually be in the tank?

c. At what exact time are there exactly 500 grams of salt present in the tank?

d. Can you determine the concentration of the solution that is being delivered by the inflow to the tank? If yes, explain why and determine this value. If not, explain why that information cannot be found without additional data.

CHAPTER 4

Trigonometry

4.1 Right triangles

Motivating Questions

- How can we view $\cos(\theta)$ and $\sin(\theta)$ as side lengths in right triangles with hypotenuse 1?

- Why can both $\cos(\theta)$ and $\sin(\theta)$ be thought of as ratios of certain side lengths in any right triangle?

- What is the minimum amount of information we need about a right triangle in order to completely determine all of its sides and angles?

In Section 2.3, we defined the cosine and sine functions as the functions that track the location of a point traversing the unit circle counterclockwise from $(1,0)$. In particular, for a central angle of radian measure t that passes through the point $(1,0)$, we define $\cos(t)$ as the x-coordinate of the point where the other side of the angle intersects the unit circle, and $\sin(t)$ as the y-coordinate of that same point, as pictured in Figure 4.1.1.

By changing our perspective slightly, we can see that it is equivalent to think of the values of the sine and cosine function as representing the lengths of legs in right triangles. Specifically, given a central angle[1] θ, if we think of the right triangle with vertices $(\cos(\theta), 0)$, $(0,0)$, and $(\cos(\theta), \sin(\theta))$, then the length of the horizontal leg is $\cos(\theta)$ and the length of the vertical leg is $\sin(\theta)$, as seen in Figure 4.1.2.

---

[1]In our work with right triangles, we'll often represent the angle by θ and think of this angle as fixed, as opposed to our previous use of t where we frequently think of t as changing.

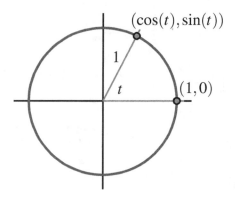

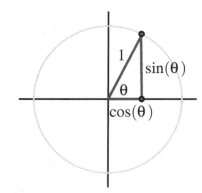

Figure 4.1.1: The values of $\cos(t)$ and $\sin(t)$ as coordinates on the unit circle.

Figure 4.1.2: The values of $\cos(\theta)$ and $\sin(\theta)$ as the lengths of the legs of a right triangle.

This right triangle perspective enables us to use the sine and cosine functions to determine missing information in certain right triangles. The field of mathematics that studies relationships among the angles and sides of triangles is called *trigonometry*. In addition, it's important to recall both the Pythagorean Theorem and the Fundamental Trigonometric Identity.

The former states that in any right triangle with legs of length a and b and hypotenuse of length c, it follows $a^2 + b^2 = c^2$. The latter, which is a special case of the Pythagorean Theorem, says that for any angle θ, $\cos^2(\theta) + \sin^2(\theta) = 1$.

Preview Activity 4.1.1. For each of the following situations, sketch a right triangle that satisfies the given conditions, and then either determine the requested missing information in the triangle or explain why you don't have enough information to determine it. Assume that all angles are being considered in radian measure.

a. The length of the other leg of a right triangle with hypotenuse of length 1 and one leg of length $\frac{3}{5}$.

b. The lengths of the two legs in a right triangle with hypotenuse of length 1 where one of the non-right angles measures $\frac{\pi}{3}$.

c. The length of the other leg of a right triangle with hypotenuse of length 7 and one leg of length 6.

d. The lengths of the two legs in a right triangle with hypotenuse 5 and where one of the non-right angles measures $\frac{\pi}{4}$.

e. The length of the other leg of a right triangle with hypotenuse of length 1 and one leg of length $\cos(0.7)$.

f. The measures of the two angles in a right triangle with hypotenuse of length 1 where the two legs have lengths $\cos(1.1)$ and $\sin(1.1)$, respectively.

4.1.1 The geometry of triangles

In the study of functions, linear functions are the simplest of all and form a foundation for our understanding of functions that have other shapes. In the study of geometric shapes (polygons, circles, and more), the simplest figure of all is the triangle, and understanding triangles is foundational to understanding many other geometric ideas. To begin, we list some familiar and important facts about triangles.

- Any triangle has 6 important features: 3 sides and 3 angles.

- In any triangle in the Cartesian plane, the sum of the measures of the interior angles is π radians (or equivalently, 180°).

- In any triangle in the plane, knowing three of the six features of a triangle is often enough information to determine the missing three features.[2]

The situation is especially nice for right triangles, because then we only have five unknown features since one of the angles is $\frac{\pi}{2}$ radians (or 90°), as demonstrated in Figure 4.1.3. If we know one of the two non-right angles, then we know the other as well. Moreover, if we know any two sides, we can immediately deduce the third, because of the Pythagorean Theorem. As we saw in Preview Activity 4.1.1, the cosine and sine functions offer additional help in determining missing information in right triangles. Indeed, while the functions $\cos(t)$ and $\sin(t)$ have many important applications in modeling periodic phenomena such as osciallating masses on springs, they also find powerful application in settings involving right triangles, such as in navigation and surveying.

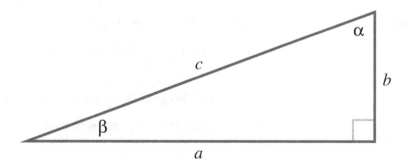

Figure 4.1.3: The 5 potential unknowns in a right triangle.

Because we know the values of the cosine and sine functions from the unit circle, right triangles with hypotentuse 1 are the easiest ones in which to determine missing information. In addition, we can relate any other right triangle to a right triangle with hypotenuse 1 through the concept of *similarity*. Recall that two triangles are **similar** provided that one is a magnification of the other. More precisely, two triangles are similar whenever there is some

---

[2]Formally, this idea relies on what are called *congruence criteria*. For instance, if we know the lengths of all three sides, then the angle measures of the triangle are uniquely determined. This is called the Side-Side-Side Criterion (SSS). You are likely familiar with SSS, as well as SAS (Side-Angle-Side), ASA, and AAS, which are the four standard criteria.

constant k such that every side in one triangle is k times as long as the corresponding side in the other and the corresponding angles in the two triangles are equal. An important result from geometry tells us that if two triangles are known to have all three of their corresponding angles equal, then it follows that the two triangles are similar, and therefore their corresponding sides must be proportionate to one another.

Activity 4.1.2. Consider right triangle OPQ given in Figure 4.1.4, and assume that the length of the hypotenuse is $OP = r$ for some constant $r > 1$. Let point M lie on $\overline{OP}$ between O and P in such a way that $OM = 1$, and let point N lie on $\overline{OQ}$ so that $\angle ONM$ is a right angle, as pictured. In addition, assume that point O corresponds to $(0,0)$, point Q to $(x,0)$, and point P to (x,y) so that $OQ = x$ and $PQ = y$. Finally, let θ be the measure of $\angle POQ$.

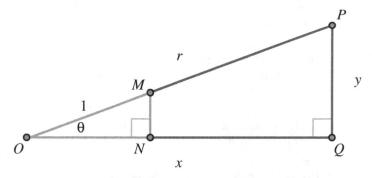

Figure 4.1.4: Two right triangles $\triangle OPQ$ and $\triangle OMN$.

a. Explain why $\triangle OPQ$ and $\triangle OMN$ are similar triangles.

b. What is the value of the ratio $\frac{OP}{OM}$? What does this tell you about the ratios $\frac{OQ}{ON}$ and $\frac{PQ}{MN}$?

c. What is the value of ON in terms of θ? What is the value of MN in terms of θ?

d. Use your conclusions in (b) and (c) to express the values of x and y in terms of r and θ.

4.1.2 Ratios of sides in right triangles

A right triangle with a hypotenuse of length 1 can be viewed as lying in standard position in the unit circle, with one vertex at the origin and one leg along the positive x-axis. If we let the angle formed by the hypotenuse and the horizontal leg have measure θ, then the right triangle with hypotenuse 1 has horizontal leg of length $\cos(\theta)$ and vertical leg of length $\sin(\theta)$. If we consider now consider a similar right triangle with hypotenuse of length $r \neq 1$, we can view that triangle as a magnification of a triangle with hypotenuse 1. These observations, combined with our work in Activity 4.1.2, show us that the horizontal legs of the right triangle with hypotenuse r have measure $r\cos(\theta)$ and $r\sin(\theta)$, as pictured in

Figure 4.1.5.

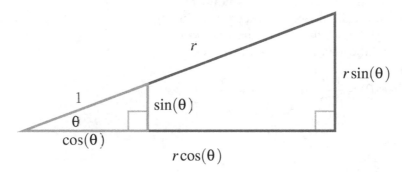

Figure 4.1.5: The roles of r and θ in a right triangle.

From the similar triangles in Figure 4.1.5, we can make an important observation about ratios in right triangles. Because the triangles are similar, the ratios of corresponding sides must be equal, so if we consider the two hypotenuses and the two horizontal legs, we have

$$\frac{r}{1} = \frac{r\cos(\theta)}{\cos(\theta)}. \tag{4.1.1}$$

If we rearrange Equation (4.1.1) by dividing both sides by r and multiplying both sides by $\cos(\theta)$, we see that

$$\frac{\cos(\theta)}{1} = \frac{r\cos(\theta)}{r}. \tag{4.1.2}$$

From a geometric perspective, Equation (4.1.2) tells us that the ratio of the horizontal leg of a right triangle to the hypotenuse of the triangle is always the same (regardless of r) and that the value of that ratio is $\cos(\theta)$, where θ is the angle adjacent to the horizontal leg. In an analogous way, the equation involving the hypotenuses and vertical legs of the similar triangles is

$$\frac{r}{1} = \frac{r\sin(\theta)}{\sin(\theta)}, \tag{4.1.3}$$

which can be rearranged to

$$\frac{\sin(\theta)}{1} = \frac{r\sin(\theta)}{r}. \tag{4.1.4}$$

Equation (4.1.4) shows that the ratio of the vertical leg of a right triangle to the hypotenuse of the triangle is always the same (regardless of r) and that the value of that ratio is $\sin(\theta)$, where θ is the angle opposite the vertical leg. We summarize these recent observations as follows.

Ratios in right triangles.

In a right triangle where one of the non-right angles is θ, and "adj" denotes the length of the leg adjacent to θ, "opp" the length the side opposite θ, and "hyp" the length of the hypotenuse,

$$\cos(\theta) = \frac{\text{adj}}{\text{hyp}} \text{ and } \sin(\theta) = \frac{\text{opp}}{\text{hyp}}.$$

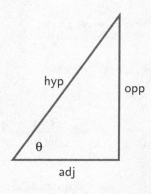

Activity 4.1.3. In each of the following scenarios involving a right triangle, determine the exact values of as many of the remaining side lengths and angle measures (in radians) that you can. If there are quantities that you cannot determine, explain why. For every prompt, draw a labeled diagram of the situation.

 a. A right triangle with hypotenuse 7 and one non-right angle of measure $\frac{\pi}{7}$.

 b. A right triangle with non-right angle α that satisfies $\sin(\alpha) = \frac{3}{5}$.

 c. A right triangle where one of the non-right angles is $\beta = 1.2$ and the hypotenuse has length 2.7.

 d. A right triangle with hypotenuse 13 and one leg of length 6.5.

 e. A right triangle with legs of length 5 and 12.

 f. A right triangle where one of the non-right angles is $\beta = \frac{\pi}{5}$ and the leg opposite this angle has length 4.

4.1.3 Using a ratio involving sine and cosine

In Activity 4.1.3, we found that in many cases where we have a right triangle, knowing two additional pieces of information enables us to find the remaining three unknown quantities in the triangle. At this point in our studies, the following general principles hold.

Missing information in right triangles.

In any right triangle,

 1. if we know one of the non-right angles and the length of the hypotenuse, we can find both the remaining non-right angle and the lengths of the two legs;

 2. if we know the length of two sides of the triangle, then we can find the length

---

of the other side;

3. if we know the measure of one non-right angle, then we can find the measure of the remaining angle.

In scenario (1.), all 6 features of the triangle are not only determined, but we are able to find their values. In (2.), the triangle is uniquely determined by the given information, but as in Activity 4.1.3 parts (d) and (e), while we know the values of the sine and cosine of the angles in the triangle, we haven't yet developed a way to determine the measures of those angles. Finally, in scenario (3.), the triangle is not uniquely determined, since any magnified version of the triangle will have the same three angles as the given one, and thus we need more information to determine side length.

We will revisit scenario (2) in our future work. Now, however, we want to consider a situation that is similar to (1), but where it is one leg of the triangle instead of the hypotenuse that is known. We encountered this in Activity 4.1.3 part (f): a right triangle where one of the non-right angles is $\beta = \frac{\pi}{5}$ and the leg opposite this angle has length 4.

Example 4.1.6

Consider a right triangle in which one of the non-right angles is $\beta = \frac{\pi}{5}$ and the leg opposite β has length 4.
Determine (both exactly and approximately) the measures of all of the remaining sides and angles in the triangle.

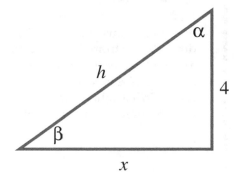

Figure 4.1.7: The given right triangle.

Solution. From the fact that $\beta = \frac{\pi}{5}$, it follows that $\alpha = \frac{\pi}{2} - \frac{\pi}{5} = \frac{3\pi}{10}$. In addition, we know that

$$\sin\left(\frac{\pi}{5}\right) = \frac{4}{h} \tag{4.1.5}$$

and

$$\cos\left(\frac{\pi}{5}\right) = \frac{x}{h} \tag{4.1.6}$$

Solving Equation (4.1.5) for h, we see that

$$h = \frac{4}{\sin\left(\frac{\pi}{5}\right)}, \tag{4.1.7}$$

which is the exact numerical value of h. Substituting this result in Equation (4.1.6), solving for h we find that

$$\cos\left(\frac{\pi}{5}\right) = \frac{x}{\frac{4}{\sin(\frac{\pi}{5})}}. \tag{4.1.8}$$

Solving this equation for the single unknown x shows that

$$x = \frac{4\cos\left(\frac{\pi}{5}\right)}{\sin\left(\frac{\pi}{5}\right)}.$$

The approximate values of x and h are $x \approx 5.506$ and $h \approx 6.805$. □

Example 4.1.6 demonstrates that a ratio of values of the sine and cosine function can be needed in order to determine the value of one of the missing sides of a right triangle, and also that we may need to work with two unknown quantities simultaneously in order to determine both of their values.

Activity 4.1.4.

We want to determine the distance between two points A and B that are directly across from one another on opposite sides of a river, as pictured in Figure 4.1.8. We mark the locations of those points and walk 50 meters downstream from B to point P and use a sextant to measure $\angle BPA$. If the measure of $\angle BPA$ is 56.4°, how wide is the river? What other information about the situation can you determine?

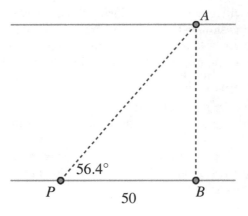

Figure 4.1.8: Finding the width of the river.

4.1.4 Summary

- In a right triangle with hypotenuse 1, we can view $\cos(\theta)$ as the length of the leg adjacent to θ and $\sin(\theta)$ as the length of the leg opposite θ, as seen in Figure 4.1.2. This is simply a change in perspective achieved by focusing on the triangle as opposed to the unit circle.

- Because a right triangle with hypotenuse of length r can be thought of as a scaled version of a right triangle with hypotenuse of length 1, we can conclude that in a right triangle with hypotenuse of length r, the leg adjacent to angle θ has length $r\cos(\theta)$, and the leg opposite θ has length $r\sin(\theta)$, as seen in Figure 4.1.5. Moreover, in any right triangle with angle θ, we know that

$$\cos(\theta) = \frac{\text{adj}}{\text{hyp}} \text{ and } \sin(\theta) = \frac{\text{opp}}{\text{hyp}}.$$

- In a right triangle, there are five additional characteristics: the measures of the two non-right angles and the lengths of the three sides. In general, if we know one of those two angles and one of the three sides, we can determine all of the remaining pieces.

4.1.5 Exercises

1. Refer to the right triangle in the figure.

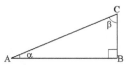

If , $BC = 3$ and the angle $\alpha = 65°$, find any missing angles or sides.

2. Suppose that a, b and c are the sides of a right triangle, where side a is across from angle A, side b is across from angle B, and side c is across from the right angle. If $a = 17$ and $B = 33°$, find the missing sides and angles in this right triangle. All angles should be in degrees (not radians), and all trig functions entered will be evaluated in degrees (not radians).

3. A person standing 50 feet away from a streetlight observes that they cast a shadow that is 14 feet long. If a ray of light from the streetlight to the tip of the person's shadow forms an angle of 27.5° with the ground, how tall is the person and how tall is the streetlight? What other information about the situation can you determine?

4. A person watching a rocket launch uses a laser range-finder to measure the distance from themselves to the rocket. The range-finder also reports the angle at which the finder is being elevated from horizontal. At a certain instant, the range-finder reports that it is elevated at an angle of 17.4° from horizontal and that the distance to the rocket is 1650 meters. How high off the ground is the rocket? Assuming a straight-line vertical path for the rocket that is perpendicular to the earth, how far away was the rocket from the range-finder at the moment it was launched?

5. A trough is constructed by bending a 4′×24′ rectangular sheet of metal. Two symmetric folds 2 feet apart are made parallel to the longest side of the rectangle so that the trough has cross-sections in the shape of a trapezoid, as pictured in Figure 4.1.9. Determine a formula for $V(\theta)$, the volume of the trough as a function of θ.

Figure 4.1.9: A cross-section of the trough.

Hint. The volume of the trough is the area of a cross-section times the length of the trough.

4.2 The Tangent Function

Motivating Questions

- How is the tangent function defined in terms of the sine and cosine functions?

- Why is the graph of the tangent function so different from the graphs of the sine and cosine functions?

- What are important applications of the tangent function?

In Activity 4.1.4, we determined the distance between two points A and B on opposite sides of a river by knowing a length along one shore of the river and the angle formed between a point downstream and the point on the opposite shore, as pictured in Figure 4.2.1. By first using the cosine of the angle, we determined the value of z and from there were able to use the sine of the angle to find w, the river's width, which turns out to be

$$w = 50 \cdot \frac{\sin(56.4)}{\cos(56.4)}.$$

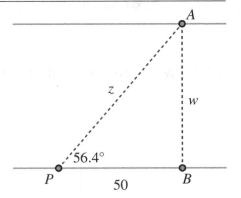

Figure 4.2.1: Finding the width of the river.

It turns out that we regularly need to evaluate the ratio of the sine and cosine functions at the same angle, so it is convenient to define a new function to be their ratio.

Definition 4.2.2 The tangent function. For any real number t for which $\cos(t) \neq 0$, we define the **tangent of** t, denoted $\tan(t)$, by

$$\tan(t) = \frac{\sin(t)}{\cos(t)}.$$

◊

Preview Activity 4.2.1. Through the following questions, we work to understand the special values and overall behavior of the tangent function.

a. Without using computational device, find the exact value of the $\tan(t)$ at the following values: $t = \frac{\pi}{6}, \frac{\pi}{4}, \frac{\pi}{3}, \frac{2\pi}{3}, \frac{3\pi}{4}, \frac{5\pi}{6}$.

b. Why is $\tan\left(\frac{\pi}{2}\right)$ not defined? What are three other input values x for which $\tan(x)$ is not defined?

c. Point your browser to http://gvsu.edu/s/0yO ("zero-y-Oh") to find a *Desmos* worksheet with data from the tangent function already input. Click on several of the orange points to compare your exact values in (a) with the decimal values given by *Desmos*. Add one entry to the table: $x = \frac{11\pi}{24}$, $y = T(\frac{11\pi}{24})$. At about what coordinates does this point lie? What are the respective values of $\sin(\frac{11\pi}{24})$ and $\cos(\frac{11\pi}{24})$? Why is the value of $\tan(\frac{11\pi}{24})$ so large?

d. At the top of the input lists on the left side of the *Desmos* worksheet, click the circle to highlight the function $T(x) = \tan(x)$ and thus show its plot along with the data points in orange. Use the plot and your work above to answer the following important questions about the tangent function:

   - What is the domain of $y = \tan(x)$?
   - What is the period of $y = \tan(x)$?
   - What is the range of $y = \tan(x)$?

4.2.1 Two perspectives on the tangent function

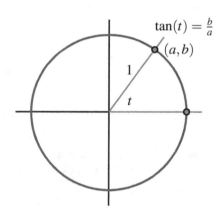

Figure 4.2.3: An angle t in standard position in the unit circle that intercepts an arc from $(1,0)$ to (a,b).

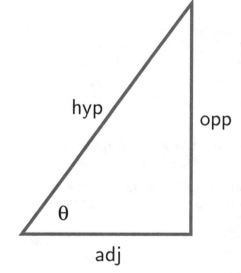

Figure 4.2.4: A right triangle with legs adjacent and opposite angle θ.

Because the tangent function is defined in terms of the two fundamental circular functions by the rule $\tan(t) = \frac{\sin(t)}{\cos(t)}$, we can use our understanding of the sine and cosine functions to make sense of the tangent function. In particular, we can think of the tangent of an angle from two different perspectives: as an angle in standard position in the unit circle, or as an angle in a right triangle.

From the viewpoint of Figure 4.2.3, as the point corresponding to angle t traverses the circle and generates the point (a,b), we know $\cos(t) = a$ and $\sin(t) = b$, and therefore the tangent function tracks the ratio of these two quantities, and is given by

$$\tan(t) = \frac{\sin(t)}{\cos(t)} = \frac{b}{a}.$$

From the perspective of any right triangle (not necessarily in the unit circle) with hypotenuse "hyp" and legs "adj" and "opp" that are respectively adjacent and opposite the known angle θ, as seen in Figure 4.2.4, we know that $\sin(\theta) = \frac{\text{opp}}{\text{hyp}}$ and $\cos(\theta) = \frac{\text{adj}}{\text{hyp}}$. Substituting these expressions for $\sin(\theta)$ and $\cos(\theta)$ in the rule for the tangent function, we see that

$$\tan(\theta) = \frac{\sin(\theta)}{\cos(\theta)} = \frac{\frac{\text{opp}}{\text{hyp}}}{\frac{\text{adj}}{\text{hyp}}} = \frac{\text{opp}}{\text{adj}}.$$

We typically use the first perspective of tracking the ratio of the y-coordinate to the x-coordinate of a point traversing the unit circle in order to think of the overall behavior and graph of the tangent function, and use the second perspective in a right triangle whenever we are working to determine missing values in a triangle.

4.2.2 Properties of the tangent function

Because the tangent function is defined in terms of the sine and cosine functions, its values and behavior are completely determined by those two functions. To begin, we know the value of $\tan(t)$ for every special angle t on the unit circle that we identified for the sine and cosine functions. For instance, we know that

$$\tan\left(\frac{\pi}{6}\right) = \frac{\sin\left(\frac{\pi}{6}\right)}{\cos\left(\frac{\pi}{6}\right)} = \frac{\frac{1}{2}}{\frac{\sqrt{3}}{2}} = \frac{1}{\sqrt{3}}.$$

Executing similar computations for every familiar special angle on the unit circle, we find the results shown in Table 4.2.5 and Table 4.2.6. We also note that anywhere $\cos(t) = 0$, the value of $\tan(t)$ is undefined. We record such instances in the table by writing "u". Table 4.2.5 and Table 4.2.6 helps us identify trends in the tangent function. For instance, we observe that the sign of $\tan(t)$ is positive in Quadrant I, negative in Quadrant II, positive in Quadrant III, and negative in Quadrant IV. This holds because the sine and cosine functions have the same sign in the first and third quadrants, and opposite signs in the other two quadrants.

In addition, we observe that as t-values in the first quadrant get closer to $\frac{\pi}{2}$, $\sin(t)$ gets closer to 1, while $\cos(t)$ gets closer to 0 (while being always positive). Noting that $\frac{\pi}{2} \approx 1.57$, we observe that

$$\tan(1.47) = \frac{\sin(1.47)}{\cos(1.47)} \approx \frac{0.995}{0.101} = 9.887$$

t	0	$\frac{\pi}{6}$	$\frac{\pi}{4}$	$\frac{\pi}{3}$	$\frac{\pi}{2}$	$\frac{2\pi}{3}$	$\frac{3\pi}{4}$	$\frac{5\pi}{6}$	π
$\sin(t)$	0	$\frac{1}{2}$	$\frac{\sqrt{2}}{2}$	$\frac{\sqrt{3}}{2}$	1	$\frac{\sqrt{3}}{2}$	$\frac{\sqrt{2}}{2}$	$\frac{1}{2}$	0
$\cos(t)$	1	$\frac{\sqrt{3}}{2}$	$\frac{\sqrt{2}}{2}$	$\frac{1}{2}$	0	$-\frac{1}{2}$	$-\frac{\sqrt{2}}{2}$	$-\frac{\sqrt{3}}{2}$	-1
$\tan(t)$	0	$\frac{1}{\sqrt{3}}$	1	$\frac{3}{\sqrt{3}}$	u	$-\frac{3}{\sqrt{3}}$	-1	$-\frac{1}{\sqrt{3}}$	0

Table 4.2.5: Values of the sine, cosine, and tangent functions at special points on the unit circle.

t	$\frac{7\pi}{6}$	$\frac{5\pi}{4}$	$\frac{4\pi}{3}$	$\frac{3\pi}{2}$	$\frac{5\pi}{3}$	$\frac{7\pi}{4}$	$\frac{11\pi}{6}$	2π
$\sin(t)$	$-\frac{1}{2}$	$-\frac{\sqrt{2}}{2}$	$-\frac{\sqrt{3}}{2}$	-1	$-\frac{\sqrt{3}}{2}$	$-\frac{\sqrt{2}}{2}$	$-\frac{1}{2}$	0
$\cos(t)$	$-\frac{\sqrt{3}}{2}$	$-\frac{\sqrt{2}}{2}$	$-\frac{1}{2}$	0	$\frac{1}{2}$	$\frac{\sqrt{2}}{2}$	$\frac{\sqrt{3}}{2}$	0
$\tan(t)$	$\frac{1}{\sqrt{3}}$	1	$\frac{3}{\sqrt{3}}$	u	$-\frac{3}{\sqrt{3}}$	-1	$-\frac{1}{\sqrt{3}}$	0

Table 4.2.6: Additional values of the sine, cosine, and tangent functions at special points on the unit circle.

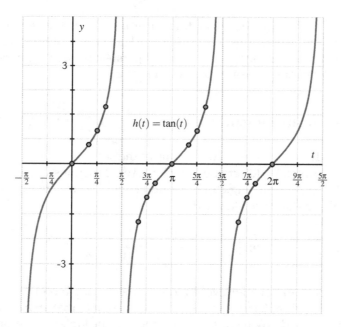

Figure 4.2.7: A plot of the tangent function together with special points that come from the unit circle.

and

$$\tan(1.56) = \frac{\sin(1.56)}{\cos(1.56)} \approx \frac{0.9994}{0.0108} = 92.6205.$$

Because the ratio of numbers closer and closer to 1 divided by numbers closer and closer to 0 (but positive) increases without bound, this means that $\tan(t)$ increases without bound as t approaches $\frac{\pi}{2}$ from the left side. Once t is slightly greater than $\frac{\pi}{2}$ in Quadrant II, the value of $\sin(t)$ stays close to 1, but now the value of $\cos(t)$ is negative (and close to zero). For instance, $\cos(1.58) \approx -0.0092$. This makes the value of $\tan(t)$ decrease without bound (negative and getting further away from 0) for t approaching $\frac{\pi}{2}$ from the right side, and results in $h(t) = \tan(t)$ having a vertical asymptote at $t = \frac{\pi}{2}$. The periodicity and sign behaviors of $\sin(t)$ and $\cos(t)$ mean this asymptotic behavior of the tangent function will repeat.

Plotting the data in the table along with the expected asymptotes and connecting the points intuitively, we see the graph of the tangent function in Figure 4.2.7. We see from Table 4.2.5

and Table 4.2.6 as well as from Figure 4.2.7 that the tangent function has period $P = \pi$ and that the function is increasing on any interval on which it is defined. We summarize our recent work as follows.

Properties of the tangent function.

For the function $h(t) = \tan(t)$,

- its domain is the set of all real numbers except $t = \frac{\pi}{2} \pm k\pi$ where k is any whole number;

- its range is the set of all real numbers;

- its period is $P = \pi$;

- is increasing on any interval on which the function is defined at every point in the interval.

While the tangent function is an interesting mathematical function for its own sake, its most important applications arise in the setting of right triangles, and for the remainder of this section we will focus on that perspective.

4.2.3 Using the tangent function in right triangles

The tangent function offers us an additional choice when working in right triangles with limited information. In the setting where we have a right triangle with one additional known angle, if we know the length of the hypotenuse, we can use either the sine or cosine of the angle to help us easily find the remaining side lengths. But in the setting where we know only the length of one leg, the tangent function now allows us to determine the value of the remaining leg in a similarly straightforward way, and from there the hypotenuse.

Example 4.2.8 Use the tangent function to determine the width, w, of the river in Figure 4.2.9. (Note that here we are revisiting the problem in Activity 4.1.4, which we previously solved without using the tangent function.) What other information can we now easily determine?

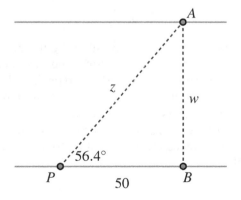

Figure 4.2.9: A right triangle with one angle and one leg known.

Solution. Using the perspective that $\tan(\theta) = \frac{\text{opp}}{\text{adj}}$ in a right triangle, in this context we have

$$\tan(56.4°) = \frac{w}{50}$$

and thus $w = 50\tan(56.4)$ is the exact width of the river. Using a computational device, we find that $w \approx 75.256$.

Once we know the river's width, we can use the Pythagorean theorem or the sine function to determine the distance from P to A, at which point all 6 parts of the triangle are known.

□

The tangent function finds a wide range of applications in finding missing information in right triangles where information about one or more legs of the triangle is known.

Activity 4.2.2. The top of a 225 foot tower is to be anchored by four cables that each make an angle of 32.5° with the ground. How long do the cables have to be and how far from the base of the tower must they be anchored?

Activity 4.2.3.

Supertall[1] high rises have changed the Manhattan skyline. These skyscrapers are known for their small footprint in proportion to their height, with their ratio of width to height at most 1 : 10, and some as extreme as 1 : 24. Suppose that a relatively short supertall has been built to a height of 635 feet, as pictured in Figure 4.2.10, and that a second supertall is built nearby. Given the two angles that are computed from the new building, how tall, s, is the new building, and how far apart, d, are the two towers?

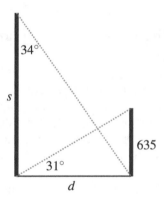

Figure 4.2.10: Two supertall skyscrapers.

Activity 4.2.4. Surveyors are trying to determine the height of a hill relative to sea level. First, they choose a point to take an initial measurement with a sextant that shows the angle of elevation from the ground to the peak of the hill is 19°. Next, they move 1000 feet closer to the hill, staying at the same elevation relative to sea level, and find that the angle of elevation has increased to 25°, as pictured in Figure 4.2.11. We let h represent the height of the hill relative to the two measurements, and x represent the distance from the second measurement location to the "center" of the hill that lies directly under the peak.

---

[1]See, for instance, this article

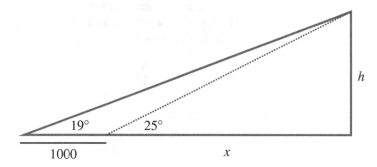

Figure 4.2.11: The surveyors' initial measurements.

a. Using the right triangle with the 25° angle, find an equation that relates x and h.

b. Using the right triangle with the 19° angle, find a second equation that relates x and h.

c. Our work in (a) and (b) results in a system of two equations in the two unknowns x and h. Solve each of the two equations for h and then substitute appropriately in order to find a single equation in the variable x.

d. Solve the equation from (c) to find the exact value of x and determine an approximate value accurate to 3 decimal places.

e. Use your preceding work to solve for h exactly, plus determine an estimate accurate to 3 decimal places.

f. If the surveyors' initial measurements were taken from an elevation of 78 feet above sea level, how high above sea level is the peak of the hill?

4.2.4 Summary

- The tangent function is defined defined to be the ratio of the sine and cosine functions according to the rule

$$\tan(t) = \frac{\sin(t)}{\cos(t)}$$

for all values of t for which $\cos(t) \neq 0$.

- The graph of the tangent function differs substantially from the graphs of the sine and cosine functions, primarily because near values where $\cos(t) = 0$, the ratio of $\frac{\sin(t)}{\cos(t)}$ increases or decreases without bound, producing vertical asymptotes. In addition, while the period of the sine and cosine functions is $P = 2\pi$, the period of the tangent function is $P = \pi$ due to how the sine and cosine functions repeat the same values (with different signs) as a point traverses the unit circle.

- The tangent function finds some of its most important applications in the setting of right triangles where one leg of the triangle is known and one of the non-right angles is known. Computing the tangent of the known angle, say α, and using the fact that

$$\tan(\alpha) = \frac{\text{opp}}{\text{adj}}$$

we can then find the missing leg's length in terms of the other and the tangent of the angle.

4.2.5 Exercises

1. From the information given, find the quadrant in which the terminal point determined by t lies. Input I, II, III, or IV.

(a) $\sin(t) < 0$ and $\cos(t) < 0$

(b) $\sin(t) > 0$ and $\cos(t) < 0$

(c) $\sin(t) > 0$ and $\cos(t) > 0$

(d) $\sin(t) < 0$ and $\cos(t) > 0$

2. Refer to the right triangle in the figure.

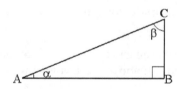

If $BC = 6$ and the angle $\alpha = 65°$, find any missing angles or sides. Give your answer to at least 3 decimal digits.

3. Refer to the right triangle in the figure.

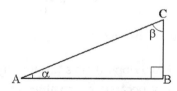

If $AC = 7$ and the angle $\beta = 45°$, find any missing angles or sides. Give your answer to at least 3 decimal digits.

4. If $\cos(\phi) = 0.8347$ and $3\pi/2 \leq \phi \leq 2\pi$, approximate the following to four decimal places.

   (a) $\sin(\phi)$

   (b) $\tan(\phi)$

5. Suppose $\sin\theta = \dfrac{x}{8}$ and the angle θ is in the first quadrant. Write algebraic expressions for $\cos(\theta)$ and $\tan(\theta)$ in terms of x.

   (a) $\cos(\theta)$

   (b) $\tan(\theta)$

6. Solve the equations below exactly. Give your answers in radians, and find all possible values for t in the interval $0 \leq t \leq 2\pi$.

   (a) $\sin(t) = \dfrac{\sqrt{2}}{2}$

   (b) $\cos(t) = -\dfrac{\sqrt{2}}{2}$

   (c) $\tan(t) = -\dfrac{1}{\sqrt{3}}$

7. A wheelchair ramp is to be built so that the angle it forms with level ground is $4°$. If the ramp is going to rise from a level sidewalk up to a front porch that is 3 feet above the ground, how long does the ramp have to be? How far from the front porch will it meet the sidewalk? What is the slope of the ramp?

8. A person is flying a kite and at the end of a fixed length of string. Assume there is no slack in the string.

   At a certain moment, the kite is 170 feet off the ground, and the angle of elevation the string makes with the ground is $40°$.

   a. How far is it from the person flying the kite to another person who is standing directly beneath the kite?

   b. How much string is out between the person flying the kite and the kite itself?

   c. With the same amount of string out, the angle of elevation increases to $50°$. How high is the kite at this time?

9. An airplane is flying at a constant speed along a straight path above a straight road at a constant elevation of 2400 feet. A person on the road observes the plane flying directly at them and uses a sextant to measure the angle of elevation from them to the plane. The first measurement they take records an angle of $36°$; a second measurement taken 2 seconds later is $41°$.

   How far did the plane travel during the two seconds between the two angle measurements? How fast was the plane flying?

4.3 Inverses of trigonometric functions

Motivating Questions

- Is it possible for a periodic function that fails the Horizontal Line Test to have an inverse?

- For the restricted cosine, sine, and tangent functions, how do we define the corresponding arccosine, arcsine, and arctangent functions?

- What are the key properties of the arccosine, arcsine, and arctangent functions?

In our prior work with inverse functions, we have seen several important principles, including

- A function f has an inverse function if and only if there exists a function g that undoes the work of f. Such a function g has the properties that $g(f(x)) = x$ for each x in the domain of f, and $f(g(y)) = y$ for each y in the range of f. We call g the inverse of f, and write $g = f^{-1}$.

- A function f has an inverse function if and only if the graph of f passes the Horizontal Line Test.

- When f has an inverse, we know that writing "$y = f(t)$" and "$t = f^{-1}(y)$" say the exact same thing, but from two different perspectives.

The trigonometric functions $f(t) = \sin(t)$, $g(t) = \cos(t)$, and $h(t) = \tan(t)$ are periodic, so each fails the horizontal line test, and thus these functions on their full domains do not have inverse functions. At the same time, it is reasonable to think about changing perspective and viewing angles as outputs in certain restricted settings. For instance, we may want to say both

$$\frac{\sqrt{3}}{2} = \cos\left(\frac{\pi}{6}\right) \quad \text{and} \quad \frac{\pi}{6} = \cos^{-1}\left(\frac{\sqrt{3}}{2}\right)$$

depending on the context in which we are considering the relationship between the angle and side length.

It's also important to understand why the issue of finding an angle in terms of a known value of a trigonometric function is important. Suppose we know the following information about a right triangle: one leg has length 2.5, and the hypotenuse has length 4. If we let θ be the angle opposite the side of length 2.5, it follows that $\sin(\theta) = \frac{2.5}{4}$. We naturally want to use the inverse of the sine function to solve the most recent equation for θ. But the sine function does not have an inverse function, so how can we address this situation?

While the original trigonometric functions $f(t) = \sin(t)$, $g(t) = \cos(t)$, and $h(t) = \tan(t)$ do not have inverse functions, it turns out that we can consider restricted versions of them that do have corresponding inverse functions. We thus investigate how we can think differently about the trigonometric functions so that we can discuss inverses in a meaningful way.

Preview Activity 4.3.1. Consider the plot of the standard cosine function in Figure 4.3.1 along with the emphasized portion of the graph on $[0, \pi]$.

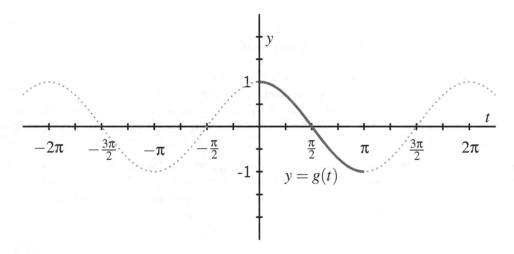

Figure 4.3.1: The cosine function on $[-\frac{5\pi}{2}, \frac{5\pi}{2}]$ with the portion on $[0, \pi]$ emphasized.

Let g be the function whose domain is $0 \leq t \leq \pi$ and whose outputs are determined by the rule $g(t) = \cos(t)$. *Note well*: g is defined in terms of the cosine function, but because it has a different domain, it is *not* the cosine function.

a. What is the domain of g?

b. What is the range of g?

c. Does g pass the horizontal line test? Why or why not?

d. Explain why g has an inverse function, g^{-1}, and state the domain and range of g^{-1}.

e. We know that $g(\frac{\pi}{4}) = \frac{\sqrt{2}}{2}$. What is the exact value of $g^{-1}(\frac{\sqrt{2}}{2})$? How about the exact value of $g^{-1}(-\frac{\sqrt{2}}{2})$?

f. Determine the exact values of $g^{-1}(-\frac{1}{2})$, $g^{-1}(\frac{\sqrt{3}}{2})$, $g^{-1}(0)$, and $g^{-1}(-1)$. Use proper notation to label your results.

4.3.1 The arccosine function

For the cosine function restricted to the domain $[0, \pi]$ that we considered in Preview Activity 4.3.1, the function is strictly decreasing on its domain and thus passes the Horizontal Line Test. Therefore, this restricted version of the cosine function has an inverse function;

we will call this inverse function the **arccosine** function.

Definition 4.3.2 Let $y = g(t) = \cos(t)$ be defined on the domain $[0, \pi]$, and observe $g : [0, \pi] \rightarrow [-1, 1]$. For any real number y that satisfies $-1 \leq y \leq 1$, the **arccosine of** y, denoted

$$\arccos(y)$$

is the angle t satisfying $0 \leq t \leq \pi$ such that $\cos(t) = y$. ◊

Note particularly that the output of the arccosine function is an angle. In addition, recall that in the context of the unit circle, an angle measured in radians and the corresponding arc length along the unit circle are numerically equal. This is why we use the "arc" in "arccosine": given a value $-1 \leq y \leq 1$, the arccosine function produces the corresponding *arc* (measured counterclockwise from $(1, 0)$) such that the cosine of that arc is y.

We recall that for any function with an inverse function, the inverse function reverses the process of the original function. We know that "$y = \cos(t)$" can be read as saying "y is the cosine of the angle t". Changing perspective and writing the equivalent statement "$t = \arccos(y)$", we read this statement as "t is the angle whose cosine is y". Just as $y = f(t)$ and $t = f^{-1}(y)$ say the same thing for a function and its inverse in general,

$$y = \cos(t) \quad \text{and} \quad t = \arccos(y)$$

say the same thing for any angle t that satisfies $0 \leq t \leq \pi$. We also use the equivalent notation $t = \cos^{-1}(y)$ interchangeably with $t = \arccos(y)$. We read "$t = \cos^{-1}(y)$" as "t is the angle whose cosine is y" or "t is the inverse cosine of y". Key properties of the arccosine function can be summarized as follows.

Properties of the arccosine function.

- The restricted cosine function, $y = g(t) = \cos(t)$, is defined on the domain $[0, \pi]$ with range $[-1, 1]$. This function has an inverse function that we call the arccosine function, denoted $t = g^{-1}(y) = \arccos(y)$.

- The domain of $y = g^{-1}(t) = \arccos(t)$ is $[-1, 1]$ with range $[0, \pi]$.

- The arccosine function is always decreasing on its domain.

- At right, a plot of the restricted cosine function (in light blue) and its corresponding inverse, the arccosine function (in dark blue).

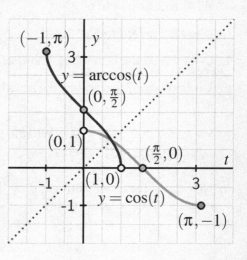

Just as the natural logarithm function allowed us to rewrite exponential equations in an equivalent way (for instance, $y = e^t$ and $t = \ln(y)$ say the exact same thing), the arccosine function allows us to do likewise for certain angles and cosine outputs. For instance, saying $\cos(\frac{\pi}{2}) = 0$ is the same as writing $\frac{\pi}{2} = \arccos(0)$, which reads "$\frac{\pi}{2}$ is the angle whose cosine is

0". Indeed, these relationships are reflected in the plot above, where we see that any point (a, b) that lies on the graph of $y = \cos(t)$ corresponds to the point (b, a) that lies on the graph of $y = \arccos(t)$.

> **Activity 4.3.2.** Use the special points on the unit circle (see, for instance, Figure 2.3.1) to determine the exact values of each of the following numerical expressions. Do so without using a computational device.
>
> a. $\arccos(\frac{1}{2})$
>
> b. $\arccos(\frac{\sqrt{2}}{2})$
>
> c. $\arccos(\frac{\sqrt{3}}{2})$
>
> d. $\arccos(-\frac{1}{2})$
>
> e. $\arccos(-\frac{\sqrt{2}}{2})$
>
> f. $\arccos(-\frac{\sqrt{3}}{2})$
>
> g. $\arccos(-1)$
>
> h. $\arccos(0)$
>
> i. $\cos(\arccos(-\frac{1}{2}))$
>
> j. $\arccos(\cos(\frac{7\pi}{6}))$

4.3.2 The arcsine function

We can develop an inverse function for a restricted version of the sine function in a similar way. As with the cosine function, we need to choose an interval on which the sine function is always increasing or always decreasing in order to have the function pass the horizontal line test. The standard choice is the domain $[-\frac{\pi}{2}, \frac{\pi}{2}]$ on which $f(t) = \sin(t)$ is increasing and attains all of the values in the range of the sine function. Thus, we consider $f(t) = \sin(t)$ so that $f : [-\frac{\pi}{2}, \frac{\pi}{2}] \to [-1, 1]$ and hence define the corresponding arcsine function.

Definition 4.3.3 Let $y = f(t) = \sin(t)$ be defined on the domain $[-\frac{\pi}{2}, \frac{\pi}{2}]$, and observe $f : [-\frac{\pi}{2}, \frac{\pi}{2}] \to [-1, 1]$. For any real number y that satisfies $-1 \leq y \leq 1$, the **arcsine of** y, denoted

$$\arcsin(y)$$

is the angle t satisfying $-\frac{\pi}{2} \leq t \leq \frac{\pi}{2}$ such that $\sin(t) = y$.   ◊

> **Activity 4.3.3.** The goal of this activity is to understand key properties of the arcsine function in a way similar to our discussion of the arccosine function in Subsection 4.3.1.
>
> a. Using Definition 4.3.3, what are the domain and range of the arcsine function?
>
> b. Determine the following values exactly: $\arcsin(-1)$, $\arcsin(-\frac{\sqrt{2}}{2})$, $\arcsin(0)$, $\arcsin(\frac{1}{2})$, and $\arcsin(\frac{\sqrt{3}}{2})$.
>
> c. On the axes provided in Figure 4.3.4, sketch a careful plot of the restricted sine function on the interval $[-\frac{\pi}{2}, \frac{\pi}{2}]$ along with its corresponding inverse, the arcsine function. Label at least three points on each curve so that each point on the sine graph corresponds to a point on the arcsine graph. In addition, sketch the line $y = t$ to demonstrate how the graphs are reflections of one another across this line.

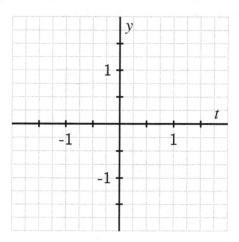

Figure 4.3.4: Axes for plotting the restricted sine function and its inverse, the arcsine function.

    d. True or false: $\arcsin(\sin(5\pi)) = 5\pi$. Write a complete sentence to explain your reasoning.

4.3.3 The arctangent function

Finally, we develop an inverse function for a restricted version of the tangent function. We choose the domain $(-\frac{\pi}{2}, \frac{\pi}{2})$ on which $h(t) = \tan(t)$ is increasing and attains all of the values in the range of the tangent function.

Definition 4.3.5 Let $y = h(t) = \tan(t)$ be defined on the domain $(-\frac{\pi}{2}, \frac{\pi}{2})$, and observe $h : (-\frac{\pi}{2}, \frac{\pi}{2}) \to (-\infty, \infty)$. For any real number y, the **arctangent of** y, denoted

$$\arctan(y)$$

is the angle t satisfying $-\frac{\pi}{2} < t < \frac{\pi}{2}$ such that $\tan(t) = y$. ◊

Activity 4.3.4. The goal of this activity is to understand key properties of the arctangent function.

    a. Using Definition 4.3.5, what are the domain and range of the arctangent function?

    b. Determine the following values exactly: $\arctan(-\sqrt{3})$, $\arctan(-1)$, $\arctan(0)$, and $\arctan(\frac{1}{\sqrt{3}})$.

    c. A plot of the restricted tangent function on the interval $(-\frac{\pi}{2}, \frac{\pi}{2})$ is provided in Figure 4.3.6. Sketch its corresponding inverse function, the arctangent function, on the same axes. Label at least three points on each curve so that each point on the tangent graph corresponds to a point on the arctangent graph. In addition, sketch the line $y = t$ to demonstrate how the graphs are reflections of one another across this line.

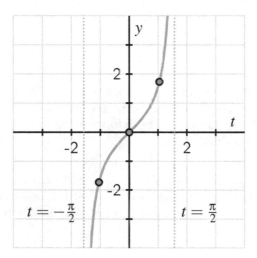

Figure 4.3.6: Axes for plotting the restricted sine function and its inverse, the arcsine function.

    d. Complete the following sentence: "as t increases without bound, $\arctan(t)\ldots$".

4.3.4 Summary

- Any function that fails the Horizontal Line Test cannot have an inverse function. However, for a periodic function that fails the horizontal line test, if we restrict the domain of the function to an interval that is the length of a single period of the function, we then determine a related function that does, in fact, have an inverse function. This makes it possible for us to develop the inverse functions of the restricted cosine, sine, and tangent functions.

- We choose to define the restricted cosine, sine, and tangent functions on the respective domains $[0, \pi]$, $[-\frac{\pi}{2}, \frac{\pi}{2}]$, and $(-\frac{\pi}{2}, \frac{\pi}{2})$. On each such interval, the restricted function is strictly decreasing (cosine) or strictly increasing (sine and tangent), and thus has an inverse function. The restricted sine and cosine functions each have range $[-1, 1]$, while the restricted tangent's range is the set of all real numbers. We thus define the inverse function of each as follows:

    i. For any y such that $-1 \leq y \leq 1$, the arccosine of y (denoted $\arccos(y)$) is the angle t in the interval $[0, \pi]$ such that $\cos(t) = y$. That is, t is the angle whose cosine is y.

ii. For any y such that $-1 \le y \le 1$, the arcsine of y (denoted $\arcsin(y)$) is the angle t in the interval $[-\frac{\pi}{2}, \frac{\pi}{2}]$ such that $\sin(t) = y$. That is, t is the angle whose sine is y.

iii. For any real number y, the arctangent of y (denoted $\arctan(y)$) is the angle t in the interval $(-\frac{\pi}{2}, \frac{\pi}{2})$ such that $\tan(t) = y$. That is, t is the angle whose tangent is y.

- To discuss the properties of the three inverse trigonometric functions, we plot them on the same axes as their corresponding restricted trigonometric functions. When we do so, we use t as the input variable for both functions simultaneously so that we can plot them on the same coordinate axes.

The domain of $y = g^{-1}(t) = \arccos(t)$ is $[-1, 1]$ with corresponding range $[0, \pi]$, and the arccosine function is always decreasing. These facts correspond to the domain and range of the restricted cosine function and the fact that the restricted cosine function is decreasing on $[0, \pi]$.

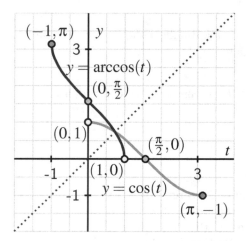

Figure 4.3.7: The restricted cosine function (in light blue) and its inverse, $y = g^{-1}(t) = \arccos(t)$ (in dark blue).

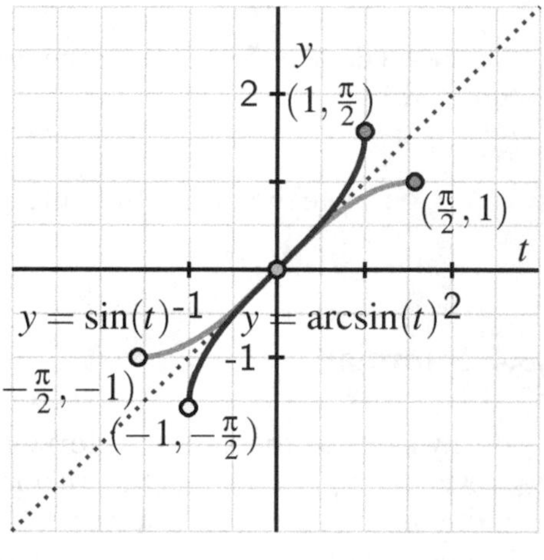

Figure 4.3.8: The restricted sine function (in light blue) and its inverse, $y = f^{-1}(t) = \arcsin(t)$ (in dark blue).

The domain of $y = f^{-1}(t) = \arcsin(t)$ is $[-1, 1]$ with corresponding range $[-\frac{\pi}{2}, \frac{\pi}{2}]$, and the arcsine function is always increasing. These facts correspond to the domain and range of the restricted sine function and the fact that the restricted sine function is increasing on $[-\frac{\pi}{2}, \frac{\pi}{2}]$.

The domain of $y = f^{-1}(t) = \arctan(t)$ is the set of all real numbers with corresponding range $(-\frac{\pi}{2}, \frac{\pi}{2})$, and the arctangent function is always increasing. These facts correspond to the domain and range of the restricted tangent function and the fact that the restricted tangent function is increasing on $(-\frac{\pi}{2}, \frac{\pi}{2})$.

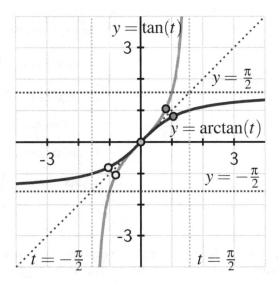

Figure 4.3.9: The restricted tangent function (in light blue) and its inverse, $y = h^{-1}(t) = \arctan(t)$ (in dark blue).

4.3.5 Exercises

1. Without using a calculator, find all solutions to $\cos(\theta) = \dfrac{\sqrt{3}}{2}$ in the interval $0 \le \theta \le 2\pi$.
Your answers should be exact values (given as fractions, not decimal approximations).

2. Without using a calculator, find all solutions to $\sin(\theta) = \dfrac{\sqrt{2}}{2}$ in the interval $0 \le \theta \le 2\pi$.
Your answers should be exact values (given as fractions, not decimal approximations).

3. Without using a calculator, find all solutions to $\tan(\theta) = 1$ in the interval $0 \le \theta \le 2\pi$.
Your answers should be exact values (given as fractions, not decimal approximations).

4. Solve the equations below exactly. Give your answers in radians, and find all possible
values for t in the interval $0 \le t \le 2\pi$.

(a) $\sin(t) = \dfrac{\sqrt{3}}{2}$ when $t =$ _____

(b) $\cos(t) = \dfrac{\sqrt{2}}{2}$ when $t =$ _____

(c) $\tan(t) = -\sqrt{3}$ when $t =$ _____

5. Use the special points on the unit circle (see, for instance, Figure 2.3.1) to determine the exact values of each of the following numerical expressions. Do so without using a computational device.

a. $\arcsin(\frac{1}{2})$

b. $\arctan(-1)$

c. $\arcsin(-\frac{\sqrt{3}}{2})$

d. $\arctan(-\frac{1}{\sqrt{3}})$

e. $\arccos(\sin(\frac{\pi}{3}))$

f. $\cos(\arcsin(-\frac{\sqrt{3}}{2}))$

g. $\tan(\arcsin(-\frac{\sqrt{2}}{2}))$

h. $\arctan(\sin(\frac{\pi}{2}))$

i. $\sin(\arcsin(-\frac{1}{2}))$

j. $\arctan(\tan(\frac{7\pi}{4}))$

6. For each of the following claims, determine whether the statement is true or false. If true, write one sentence to justify your reasoning. If false, give an example of a value that shows the claim fails.

a. For any y such that $-1 \le y \le 1$, $\sin(\arcsin(y)) = y$.

b. For any real number t, $\arcsin(\sin(t)) = t$.

c. For any real number t, $\arccos(\cos(t)) = t$.

d. For any y such that $-1 \le y \le 1$, $\cos(\arccos(y)) = y$.

e. For any real number y, $\tan(\arctan(y)) = y$.

f. For any real number t, $\arctan(\tan(t)) = t$.

7. Let's consider the composite function $h(x) = \cos(\arcsin(x))$. This function makes sense to consider since the arcsine function produces an angle, at which the cosine function can then be evaluated. In the questions that follow, we investigate how to express h without using trigonometric functions at all.

a. What is the domain of h? The range of h?

b. Since the arcsine function produces an angle, let's say that $\theta = \arcsin(x)$, so that θ is the angle whose sine is x. By definition, we can picture θ as an angle in a right triangle with hypotenuse 1 and a vertical leg of length x, as shown in Figure 4.3.10. Use the Pythagorean Theorem to determine the length of the horizontal leg as a function of x.

c. What is the value of $\cos(\theta)$ as a function of x? What have we shown about $h(x) = \cos(\arcsin(x))$?

d. How about the function $p(x) = \cos(\arctan(x))$? How can you reason similarly to write p in a way that doesn't involve any trigonometric functions at all? (Hint: let $\alpha = \arctan(x)$ and consider the right triangle in Figure 4.3.11.)

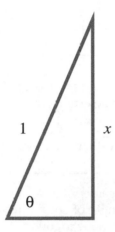

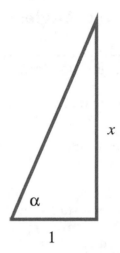

Figure 4.3.10: The right triangle that corresponds to the angle $\theta = \arcsin(x)$.

Figure 4.3.11: The right triangle that corresponds to the angle $\alpha = \arctan(x)$.

4.4 Finding Angles

Motivating Questions

- How can we use inverse trigonometric functions to determine missing angles in right triangles?

- What situations require us to use technology to evaluate inverse trignometric functions?

In our earlier work in Section 4.1 and Section 4.2, we observed that in any right triangle, if we know the measure of one additional angle and the length of one additional side, we can determine all of the other parts of the triangle. With the inverse trigonometric functions that we developed in Section 4.3, we are now also able to determine the missing angles in any right triangle where we know the lengths of two sides.

While the original trigonometric functions take a particular angle as input and provide an output that can be viewed as the ratio of two sides of a right triangle, the inverse trigonometric functions take an input that can be viewed as a ratio of two sides of a right triangle and produce the corresponding angle as output. Indeed, it's imperative to remember that statements such as

$$\arccos(x) = \theta \quad \text{and} \quad \cos(\theta) = x$$

say the exact same thing from two different perspectives, and that we read "$\arccos(x)$" as "the angle whose cosine is x".

Preview Activity 4.4.1. Consider a right triangle that has one leg of length 3 and another leg of length $\sqrt{3}$. Let θ be the angle that lies opposite the shorter leg.

   a. Sketch a labeled picture of the triangle.

   b. What is the exact length of the triangle's hypotenuse?

   c. What is the exact value of $\sin(\theta)$?

   d. Rewrite your equation from (b) using the arcsine function in the form $\arcsin(\square) = \triangle$, where $\square$ and $\triangle$ are numerical values.

   e. What special angle from the unit circle is θ?

4.4.1 Evaluating inverse trigonometric functions

Like the trigonometric functions themselves, there are a handful of important values of the inverse trigonometric functions that we can determine exactly without the aid of a computer. For instance, we know from the unit circle (Figure 2.3.1) that $\arcsin(-\frac{\sqrt{3}}{2}) = -\frac{\pi}{3}$, $\arccos(-\frac{\sqrt{3}}{2}) = \frac{5\pi}{6}$, and $\arctan(-\frac{1}{\sqrt{3}}) = -\frac{\pi}{6}$. In these evaluations, we have to be careful to remember that the range of the arccosine function is $[0, \pi]$, while the range of the arcsine function is $[-\frac{\pi}{2}, \frac{\pi}{2}]$ and the range of the arctangent function is $(-\frac{\pi}{2}, \frac{\pi}{2})$, in order to ensure that we choose the appropriate angle that results from the inverse trigonometric function.

In addition, there are many other values at which we may wish to know the angle that results from an inverse trigonometric function. To determine such values, we use a computational device (such as *Desmos*) in order to evaluate the function.

Example 4.4.1

Consider the right triangle pictured in Figure 4.4.2 and assume we know that the vertical leg has length 1 and the hypotenuse has length 3. Let α be the angle opposite the known leg. Determine exact and approximate values for all of the remaining parts of the triangle.

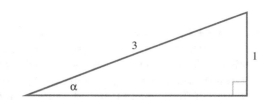

Figure 4.4.2: A right triangle with one known leg and known hypotenuse.

Solution. Because we know the hypotenuse and the side opposite α, we observe that $\sin(\alpha) = \frac{1}{3}$. Rewriting this statement using inverse function notation, we have equivalently that $\alpha = \arcsin(\frac{1}{3})$, which is the exact value of α. Since this is not one of the known special angles on the unit circle, we can find a numerical estimate of α using a computational device. Entering arcsin(1/3) in *Desmos*, we find that $\alpha \approx 0.3398$ radians. Note well: whatever device we use, we need to be careful to use degree or radian mode as dictated by the problem we are solving. We will always work in radians unless stated otherwise.

We can now find the remaining leg's length and the remaining angle's measure. If we let x represent the length of the horizontal leg, by the Pythagorean Theorem we know that

$$x^2 + 1^2 = 3^2,$$

and thus $x^2 = 8$ so $x = \sqrt{8} \approx 2.8284$. Calling the remaining angle β, since $\alpha + \beta = \frac{\pi}{2}$, it follows that

$$\beta = \frac{\pi}{2} - \arcsin\left(\frac{1}{3}\right) \approx 1.2310.$$

$\square$

> **Activity 4.4.2.** For each of the following different scenarios, draw a picture of the situation and use inverse trigonometric functions appropriately to determine the missing information both exactly and approximately.
>
> a. Consider a right triangle with legs of length 11 and 13. What are the measures (in radians) of the non-right angles and what is the length of the hypotenuse?
>
> b. Consider an angle α in standard position (vertex at the origin, one side on the positive x-axis) for which we know $\cos(\alpha) = -\frac{1}{2}$ and α lies in quadrant III. What is the measure of α in radians? In addition, what is the value of $\sin(\alpha)$?
>
> c. Consider an angle β in standard position for which we know $\sin(\beta) = 0.1$ and β lies in quadrant II. What is the measure of β in radians? In addition, what is the value of $\cos(\beta)$?

4.4.2 Finding angles in applied contexts

Now that we have developed the (restricted) sine, cosine, and tangent functions and their respective inverses, in any setting in which we have a right triangle together with one side length and any one additional piece of information (another side length or a non-right angle measurement), we can determine all of the remaining pieces of the triangle. In the activities that follow, we explore these possibilities in a variety of different applied contexts.

Activity 4.4.3. A roof is being built with a "7-12 pitch." This means that the roof rises 7 inches vertically for every 12 inches of horizontal span; in other words, the slope of the roof is $\frac{7}{12}$. What is the exact measure (in degrees) of the angle the roof makes with the horizontal? What is the approximate measure? What are the exact and approximate measures of the angle at the peak of the roof (made by the front and back portions of the roof that meet to form the ridge)?

Activity 4.4.4. On a baseball diamond (which is a square with 90-foot sides), the third baseman fields the ball right on the line from third base to home plate and 10 feet away from third base (towards home plate). When he throws the ball to first base, what angle (in degrees) does the line the ball travels make with the first base line? What angle does it make with the third base line? Draw a well-labeled diagram to support your thinking.

What angles arise if he throws the ball to second base instead?

Activity 4.4.5. A camera is tracking the launch of a SpaceX rocket. The camera is located 4000′ from the rocket's launching pad, and the camera elevates in order to keep the rocket in focus. At what angle θ (in radians) is the camera tilted when the rocket is 3000′ off the ground? Answer both exactly and approximately.

Now, rather than considering the rocket at a fixed height of 3000′, let its height vary and call the rocket's height h. Determine the camera's angle, θ as a function of h, and compute the average rate of change of θ on the intervals $[3000, 3500]$, $[5000, 5500]$, and $[7000, 7500]$. What do you observe about how the camera angle is changing?

4.4.3 Summary

- Anytime we know two side lengths in a right triangle, we can use one of the inverse trigonometric functions to determine the measure of one of the non-right angles. For instance, if we know the values of opp and adj in Figure 4.4.3, then since

$$\tan(\theta) = \frac{\text{opp}}{\text{adj}},$$

it follows that $\theta = \arctan(\frac{\text{opp}}{\text{adj}})$.

If we instead know the hypotenuse and one of the two legs, we can use either the arcsine or arccosine function accordingly.

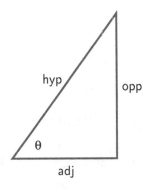

Figure 4.4.3: Finding an angle from knowing the legs in a right triangle.

- For situations other than angles or ratios that involve the 16 special points on the unit circle, technology is required in order to evaluate inverse trignometric functions. For instance, from the unit circle we know that $\arccos(\frac{1}{2}) = \frac{\pi}{3}$ (exactly), but if we want to know $\arccos(\frac{1}{3})$, we have to estimate this value using a computational device such as *Desmos*. We note that "$\arccos(\frac{1}{3})$" is the exact value of the angle whose cosine is $\frac{1}{3}$.

4.4.4 Exercises

1. If $\cos(\phi) = 0.7087$ and $3\pi/2 \le \phi \le 2\pi$, approximate the following to four decimal places.

   (a) $\sin(\phi)$

   (b) $\tan(\phi)$

2. Suppose $\sin\theta = \dfrac{x}{7}$ and the angle θ is in the first quadrant. Write algebraic expressions for $\cos(\theta)$ and $\tan(\theta)$ in terms of x.

   (a) $\cos(\theta)$

   (b) $\tan(\theta)$

3. Using inverse trigonometric functions, find a solution to the equation $\cos(x) = 0.7$ in the interval $0 \le x \le 4\pi$. Then, use a graph to find all other solutions to this equation in this interval. Enter your answers as a comma separated list.

4. At an airshow, a pilot is flying low over a runway while maintaining a constant altitude of 2000 feet and a constant speed. On a straight path over the runway, the pilot observes on her laser range-finder that the distance from the plane to a fixed building adjacent to the runway is 7500 feet. Five seconds later, she observes that distance to the same building is now 6000 feet.

   a. What is the angle of depression from the plane to the building when the plane is 7500 feet away from the building? (The angle of depression is the angle that the pilot's line of sight makes with the horizontal.)

   b. What is the angle of depression when the plane is 6000 feet from the building?

    c. How far did the plane travel during the time between the two different observations?

    d. What is the plane's velocity (in miles per hour)?

5.   On a calm day, a photographer is filming a hot air balloon. When the balloon launches, the photographer is stationed 850 feet away from the balloon.

    a. When the balloon is 200 feet off the ground, what is the angle of elevation of the camera?

    b. When the balloon is 275 feet off the ground, what is the angle of elevation of the camera?

    c. Let θ represent the camera's angle of elevation when the balloon is at an arbitrary height h above the ground. Express θ as a function of h.

    d. Determine $AV_{[200,275]}$ for θ (as a function of h) and write at least one sentence to carefully explain the meaning of the value you find, including units.

6.   Consider a right triangle where the two legs measure 5 and 12 respectively and α is the angle opposite the shorter leg and β is the angle opposite the longer leg.

    a. What is the exact value of $\cos(\alpha)$?

    b. What is the exact value of $\sin(\beta)$?

    c. What is the exact value of $\tan(\beta)$? of $\tan(\alpha)$?

    d. What is the exact radian measure of α? approximate measure?

    e. What is the exact radian measure of β? approximate measure?

    f. True or false: for any two angles θ and γ such that $\theta + \gamma = \frac{\pi}{2}$ (radians), it follows that $\cos(\theta) = \sin(\gamma)$.

4.5 Other Trigonometric Functions and Identities

Motivating Questions

- What are the other 3 trigonometric functions and how are they related to the cosine, sine, and tangent functions?

- How do the graphs of the secant, cosecant, and cotangent functions behave and how do these graphs compare to the cosine, sine, and tangent functions' graphs?

- What is a trigonometric identity and why are identities important?

The sine and cosine functions, originally defined in the context of a point traversing the unit circle, are also central in right triangle trigonometry. They enable us to find missing information in right triangles in a straightforward way when we know one of the non-right angles and one of the three sides of the triangle, or two of the sides where one is the hypotenuse. In addition, we defined the tangent function in terms of the sine and cosine functions, and the tangent function offers additional options for finding missing information in right triangles. We've also seeen how the inverses of the restricted sine, cosine, and tangent functions enable us to find missing angles in a wide variety of settings involving right triangles.

One of the powerful aspects of trigonometry is that the subject offers us the opportunity to view the same idea from many different perspectives. As one example, we have observed that the functions $f(t) = \cos(t)$ and $g(t) = \sin(t + \frac{\pi}{2})$ are actually the same function; as another, for t values in the domain $(-\frac{\pi}{2}, \frac{\pi}{2})$, we know that writing $y = \tan(t)$ is the same as writing $t = \arctan(y)$. Which perspective we choose to take often depends on context and given information.

While almost every question involving trigonometry can be answered using the sine, cosine, and tangent functions, sometimes it is convenient to use three related functions that are connected to the other three possible arrangements of ratios of sides in right triangles.

Definition 4.5.1 The secant, cosecant, and cotangent functions.

- For any real number t for which $\cos(t) \neq 0$, we define the secant of t, denoted $\sec(t)$, by the rule

$$\sec(t) = \frac{1}{\cos(t)}.$$

- For any real number t for which $\sin(t) \neq 0$, we define the cosecant of t, denoted $\csc(t)$, by the rule

$$\csc(t) = \frac{1}{\sin(t)}.$$

- For any real number t for which $\sin(t) \neq 0$, we define the cotangent of t, denoted $\cot(t)$, by the rule

$$\cot(t) = \frac{\cos(t)}{\sin(t)}.$$

◊

Note particularly that like the tangent function, the secant, cosecant, and cotangent are also defined completely in terms of the sine and cosine functions. In the context of a right triangle with an angle θ, we know how to think of $\sin(\theta)$, $\cos(\theta)$, and $\tan(\theta)$ as ratios of sides of the triangle. We can now do likewise with the other trigonometric functions:

$$\sec(\theta) = \frac{1}{\cos(\theta)} = \frac{1}{\frac{adj}{hyp}} = \frac{hyp}{adj}$$

$$\csc(\theta) = \frac{1}{\sin(\theta)} = \frac{1}{\frac{opp}{hyp}} = \frac{hyp}{opp}$$

$$\cot(\theta) = \frac{\cos(\theta)}{\sin(\theta)} = \frac{\frac{adj}{hyp}}{\frac{opp}{hyp}} = \frac{adj}{opp}$$

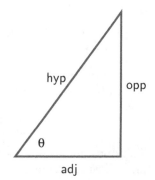

Figure 4.5.2: A right triangle with angle θ.

With these three additional trigonometric functions, we now have expressions that address all six possible combinations of two sides of a right triangle in a ratio.

Preview Activity 4.5.1. Consider a right triangle with hypotenuse of length 61 and one leg of length 11. Let α be the angle opposite the side of length 11. Find the exact length of the other leg and then determine the value of each of the six trigonometric functions evaluated at α. In addition, what are the exact and approximate measures of the two non-right angles in the triangle?

4.5.1 Ratios in right triangles

Because the sine and cosine functions are used to define each of the other four trigonometric functions, it follows that we can translate information known about the other functions back to information about the sine and cosine functions. For example, if we know that in a certain triangle $\csc(\alpha) = \frac{5}{3}$, it follows that $\sin(\alpha) = \frac{3}{5}$. From there we can reason in the usual way to determine missing information in the given triangle.

It's also often possible to view given information in the context of the unit circle. With the earlier given information that $\csc(\alpha) = \frac{5}{3}$, it's natural to view α as being the angle in a right triangle that lies opposite a leg of length 3 with the hypotenuse being 5, since $\csc(\alpha) = \frac{hyp}{opp}$. The Pythagorean Theorem then tells us the leg adjacent to α has length 4, as seen in $\triangle OPQ$ in Figure 4.5.3. But we could also view $\sin(\alpha) = \frac{3}{5}$ as $\sin(\alpha) = \frac{\frac{3}{5}}{1}$, and thus think of the right triangle has having hypotenuse 1 and vertical leg $\frac{3}{5}$. This triangle is similar to the originally considered 3-4-5 right triangle, but can be viewed as lying within the unit circle. The perspective of the unit circle is particularly valuable when ratios such as $\frac{\sqrt{3}}{2}$, $\frac{\sqrt{2}}{2}$, and $\frac{1}{2}$ arise in right triangles.

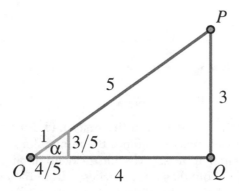

Figure 4.5.3: A 3-4-5 right triangle.

> **Activity 4.5.2.** Suppose that β is an angle in standard position with its terminal side in quadrant II and you know that $\sec(\beta) = -2$. Without using a computational device in any way, determine the exact values of the other five trigonmetric functions evaluated at β.

4.5.2 Properties of the secant, cosecant, and cotangent functions

Like the tangent function, the secant, cosecant, and cotangent functions are defined in terms of the sine and cosine functions, so we can determine the exact values of these functions at each of the special points on the unit circle. In addition, we can use our understanding of the unit circle and the properties of the sine and cosine functions to determine key properties of these other trigonometric functions. We begin by investigating the secant function.

Using the fact that $\sec(t) = \frac{1}{\cos(t)}$, we note that anywhere $\cos(t) = 0$, the value of $\sec(t)$ is undefined. We denote such instances in the following table by "u". At all other points, the value of the secant function is simply the reciprocal of the cosine function's value. Since $|\cos(t)| \leq 1$ for all t, it follows that $|\sec(t)| \geq 1$ for all t (for which the secant's value is defined). Table 4.5.4 and Table 4.5.5 help us identify trends in the secant function. The sign of $\sec(t)$ matches the sign of $\cos(t)$ and thus is positive in Quadrant I, negative in Quadrant II, negative in Quadrant III, and positive in Quadrant IV.

t	0	$\frac{\pi}{6}$	$\frac{\pi}{4}$	$\frac{\pi}{3}$	$\frac{\pi}{2}$	$\frac{2\pi}{3}$	$\frac{3\pi}{4}$	$\frac{5\pi}{6}$	π
$\cos(t)$	1	$\frac{\sqrt{3}}{2}$	$\frac{\sqrt{2}}{2}$	$\frac{1}{2}$	0	$-\frac{1}{2}$	$-\frac{\sqrt{2}}{2}$	$-\frac{\sqrt{3}}{2}$	-1
$\sec(t)$	1	$\frac{2}{\sqrt{3}}$	$\sqrt{2}$	2	u	-2	$-\sqrt{2}$	$-\frac{2}{\sqrt{3}}$	-1

Table 4.5.4: Values of the cosine and secant functions at special points on the unit circle (Quadrants I and II).

t	$\frac{7\pi}{6}$	$\frac{5\pi}{4}$	$\frac{4\pi}{3}$	$\frac{3\pi}{2}$	$\frac{5\pi}{3}$	$\frac{7\pi}{4}$	$\frac{11\pi}{6}$	2π
$\cos(t)$	$-\frac{\sqrt{3}}{2}$	$-\frac{\sqrt{2}}{2}$	$-\frac{1}{2}$	0	$\frac{1}{2}$	$\frac{\sqrt{2}}{2}$	$\frac{\sqrt{3}}{2}$	0
$\sec(t)$	$-\frac{2}{\sqrt{3}}$	$-\sqrt{2}$	-2	u	2	$\sqrt{2}$	$\frac{2}{\sqrt{3}}$	1

Table 4.5.5: Values of the cosine and secant functions at special points on the unit circle (Quadrants III and IV).

In addition, we observe that as t-values in the first quadrant get closer to $\frac{\pi}{2}$, $\cos(t)$ gets closer to 0 (while being always positive). Since the numerator of the secant function is always 1, having its denominator approach 0 (while the denominator remains positive) means that $\sec(t)$ increases without bound as t approaches $\frac{\pi}{2}$ from the left side. Once t is slightly greater than $\frac{\pi}{2}$ in Quadrant II, the value of $\cos(t)$ is negative (and close to zero). This makes the value of $\sec(t)$ decrease without bound (negative and getting further away from 0) for t approaching $\frac{\pi}{2}$ from the right side. We therefore see that $p(t) = \sec(t)$ has a vertical asymptote at $t = \frac{\pi}{2}$; the periodicity and sign behavior of $\cos(t)$ mean this asymptotic behavior of the secant function will repeat.

Plotting the data in the table along with the expected asymptotes and connecting the points intuitively, we see the graph of the secant function in Figure 4.5.6.

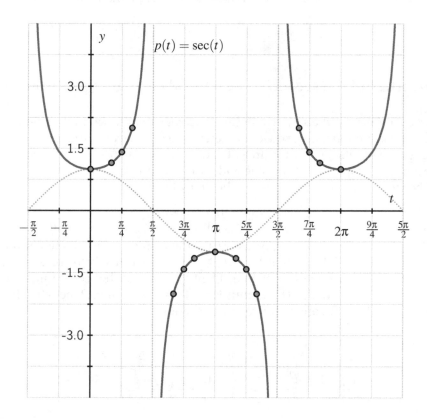

Figure 4.5.6: A plot of the secant function with special points that come from the unit circle, plus the cosine function (dotted, in light blue).

We see from both the table and the graph that the secant function has period $P = 2\pi$. We summarize our recent work as follows.

> **Properties of the secant function.**
>
> For the function $p(t) = \sec(t)$,
>
> - its domain is the set of all real numbers except $t = \frac{\pi}{2} \pm k\pi$ where k is any whole number;
>
> - its range is the set of all real numbers y such that $|y| \geq 1$;
>
> - its period is $P = 2\pi$.

Activity 4.5.3. In this activity, we develop the standard properties of the cosecant function, $q(t) = \csc(t)$.

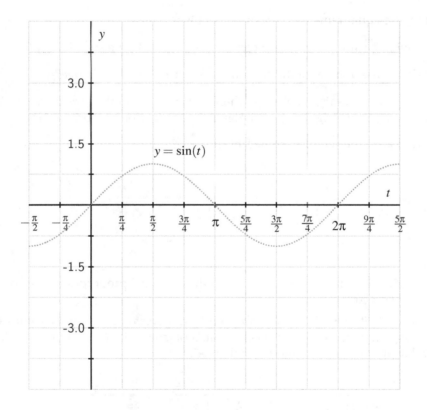

Figure 4.5.7: Axes for plotting $q(t) = \csc(t)$.

a. Complete Table 4.5.8 and Table 4.5.9 to determine the exact values of the cosecant function at the special points on the unit circle. Enter "u" for any value at which $q(t) = \csc(t)$ is undefined.

t	0	$\frac{\pi}{6}$	$\frac{\pi}{4}$	$\frac{\pi}{3}$	$\frac{\pi}{2}$	$\frac{2\pi}{3}$	$\frac{3\pi}{4}$	$\frac{5\pi}{6}$	π
$\sin(t)$	0	$\frac{1}{2}$	$\frac{\sqrt{2}}{2}$	$\frac{\sqrt{3}}{2}$	1	$\frac{\sqrt{3}}{2}$	$\frac{\sqrt{2}}{2}$	$\frac{1}{2}$	0
$\csc(t)$									

Table 4.5.8: Values of the sine function at special points on the unit circle (Quadrants I and II).

t	$\frac{7\pi}{6}$	$\frac{5\pi}{4}$	$\frac{4\pi}{3}$	$\frac{3\pi}{2}$	$\frac{5\pi}{3}$	$\frac{7\pi}{4}$	$\frac{11\pi}{6}$	2π
$\sin(t)$	$-\frac{1}{2}$	$-\frac{\sqrt{2}}{2}$	$-\frac{\sqrt{3}}{2}$	-1	$-\frac{\sqrt{3}}{2}$	$-\frac{\sqrt{2}}{2}$	$-\frac{1}{2}$	0
$\csc(t)$								

Table 4.5.9: Values of the sine function at special points on the unit circle (Quadrants III and IV).

   b. In which quadrants is $q(t) = \csc(t)$ positive? negative?

   c. At what t-values does $q(t) = \csc(t)$ have a vertical asymptote? Why?

   d. What is the domain of the cosecant function? What is its range?

   e. Sketch an accurate, labeled graph of $q(t) = \csc(t)$ on the axes provided in Figure 4.5.7, including the special points that come from the unit circle.

   f. What is the period of the cosecant function?

Activity 4.5.4. In this activity, we develop the standard properties of the cotangent function, $r(t) = \cot(t)$.

   a. Complete Table 4.5.10 and Table 4.5.11 to determine the exact values of the cotangent function at the special points on the unit circle. Enter "u" for any value at which $r(t) = \cot(t)$ is undefined.

t	0	$\frac{\pi}{6}$	$\frac{\pi}{4}$	$\frac{\pi}{3}$	$\frac{\pi}{2}$	$\frac{2\pi}{3}$	$\frac{3\pi}{4}$	$\frac{5\pi}{6}$	π
$\sin(t)$	0	$\frac{1}{2}$	$\frac{\sqrt{2}}{2}$	$\frac{\sqrt{3}}{2}$	1	$\frac{\sqrt{3}}{2}$	$\frac{\sqrt{2}}{2}$	$\frac{1}{2}$	0
$\cos(t)$	1	$\frac{\sqrt{3}}{2}$	$\frac{\sqrt{2}}{2}$	$\frac{1}{2}$	0	$-\frac{1}{2}$	$-\frac{\sqrt{2}}{2}$	$-\frac{\sqrt{3}}{2}$	-1
$\tan(t)$	0	$\frac{1}{\sqrt{3}}$	1	$\frac{3}{\sqrt{3}}$	u	$-\frac{3}{\sqrt{3}}$	-1	$-\frac{1}{\sqrt{3}}$	0
$\cot(t)$									

Table 4.5.10: Values of the sine function at special points on the unit circle.

t	$\frac{7\pi}{6}$	$\frac{5\pi}{4}$	$\frac{4\pi}{3}$	$\frac{3\pi}{2}$	$\frac{5\pi}{3}$	$\frac{7\pi}{4}$	$\frac{11\pi}{6}$	2π
$\sin(t)$	$-\frac{1}{2}$	$-\frac{\sqrt{2}}{2}$	$-\frac{\sqrt{3}}{2}$	-1	$-\frac{\sqrt{3}}{2}$	$-\frac{\sqrt{2}}{2}$	$-\frac{1}{2}$	0
$\cos(t)$	$-\frac{\sqrt{3}}{2}$	$-\frac{\sqrt{2}}{2}$	$-\frac{1}{2}$	0	$\frac{1}{2}$	$\frac{\sqrt{2}}{2}$	$\frac{\sqrt{3}}{2}$	0
$\tan(t)$	$\frac{1}{\sqrt{3}}$	1	$\frac{3}{\sqrt{3}}$	u	$-\frac{3}{\sqrt{3}}$	-1	$-\frac{1}{\sqrt{3}}$	0
$\cot(t)$								

Table 4.5.11: Values of the sine function at special points on the unit circle.

b. In which quadrants is $r(t) = \cot(t)$ positive? negative?

c. At what t-values does $r(t) = \cot(t)$ have a vertical asymptote? Why?

d. What is the domain of the cotangent function? What is its range?

e. Sketch an accurate, labeled graph of $r(t) = \cot(t)$ on the axes provided in Figure 4.5.12, including the special points that come from the unit circle.

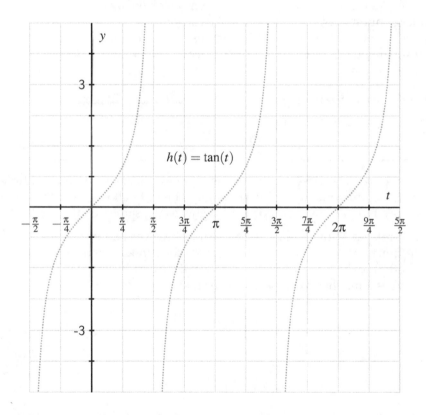

Figure 4.5.12: Axes for plotting $r(t) = \cot(t)$.

f. On intervals where the function is defined at every point in the interval, is $r(t) = \cot(t)$ always increasing, always decreasing, or neither?

g. What is the period of the cotangent function?

h. How would you describe the relationship between the graphs of the tangent and cotangent functions?

4.5.3 A few important identities

An *identity* is an equation that is true for all possible values of x for which the involved quantities are defined. An example of a non-trigonometric identity is

$$(x+1)^2 = x^2 + 2x + 1,$$

since this equation is true for every value of x, and the left and right sides of the equation are simply two different-looking but entirely equivalent expressions.

Trigonometric identities are simply identities that involve trigonometric functions. While there are a large number of such identities one can study, we choose to focus on those that turn out to be most useful in the study of calculus. The most important trigonometric identity is the fundamental trigonometric identity, which is a trigonometric restatement of the Pythagorean Theorem.

The fundamental trigonometric identity.

For any real number θ,
$$\cos^2(\theta) + \sin^2(\theta) = 1. \tag{4.5.1}$$

Identities are important because they enable us to view the same idea from multiple perspectives. For example, the fundamental trigonometric identity allows us to think of $\cos^2(\theta) + \sin^2(\theta)$ as simply 1, or alternatively, to view $\cos^2(\theta)$ as the same quantity as $1 - \sin^2(\theta)$.

There are two related Pythagorean identities that involve the tangent, secant, cotangent, and cosecant functions, which we can derive from the fundamental trigonometric identity by dividing both sides by either $\cos^2(\theta)$ or $\sin^2(\theta)$. If we divide both sides of Equation (4.5.1) by $\cos^2(\theta)$ (and assume that $\cos(\theta) \neq 0$), we see that

$$1 + \frac{\sin^2(\theta)}{\cos^2(\theta)} = \frac{1}{\cos^2(\theta)},$$

or equivalently,

$$1 + \tan^2(\theta) = \sec^2(\theta).$$

A similar argument dividing by $\sin^2(\theta)$ (while assuming $\sin(\theta) \neq 0$) shows that

$$\cot^2(\theta) + 1 = \csc^2(\theta).$$

These identities prove useful in calculus when we develop the formulas for the derivatives of the tangent and cotangent functions.

In calculus, it is also beneficial to know a couple of other standard identities for sums of angles or double angles. We simply state these identities without justification. For more information about them, see Section 10.4 in College Trigonometry, by Stitz and Zeager[1].

- For all real numbers α and β, $\cos(\alpha + \beta) = \cos(\alpha)\cos(\beta) - \sin(\alpha)\sin(\beta)$.

- For all real numbers α and β, $\sin(\alpha + \beta) = \sin(\alpha)\cos(\beta) + \cos(\alpha)\sin(\beta)$.

- For any real number θ, $\cos(2\theta) = \cos^2(\theta) - \sin^2(\theta)$.

- For any real number θ, $\sin(2\theta) = 2\sin(\theta)\cos(\theta)$.

Activity 4.5.5. In this activity, we investigate how a sum of two angles identity for the sine function helps us gain a different perspective on the average rate of change of the sine function.

Recall that for any function f on an interval $[a, a+h]$, its average rate of change is

$$AV_{[a,a+h]} = \frac{f(a+h) - f(a)}{h}.$$

a. Let $f(x) = \sin(x)$. Use the definition of $AV_{[a,a+h]}$ to write an expression for the average rate of change of the sine function on the interval $[a+h, a]$.

b. Apply the sum of two angles identity for the sine function,

$$\sin(\alpha + \beta) = \sin(\alpha)\cos(\beta) + \cos(\alpha)\sin(\beta),$$

to the expression $\sin(a + h)$.

c. Explain why your work in (a) and (b) together with some algebra shows that

$$AV_{[a,a+h]} = \sin(a) \cdot \frac{\cos(h) - 1}{h} - \cos(a)\frac{\sin(h)}{h}.$$

d. In calculus, we move from *average* rate of change to *instantaneous* rate of change by letting h approach 0 in the expression for average rate of change. Using a computational device in radian mode, investigate the behavior of

$$\frac{\cos(h) - 1}{h}$$

as h gets close to 0. What happens? Similarly, how does $\frac{\sin(h)}{h}$ behave for small values of h? What does this tell us about $AV_{[a,a+h]}$ for the sine function as h approaches 0?

---

[1]More information on Stitz and Zeager's free texts can be found at http://stitz-zeager.com/.

4.5.4 Summary

- The secant, cosecant, and cotangent functions are respectively defined as the reciprocals of the cosine, sine, and tangent functions. That is,

$$\sec(t) = \frac{1}{\cos(t)}, \csc(t) = \frac{1}{\sin(t)}, \text{ and } \cot(t) = \frac{1}{\tan(t)}.$$

- The graph of the cotangent function is similar to the graph of the tangent function, except that it is decreasing on every interval on which it is defined at every point in the interval and has vertical asymptotes wherever $\tan(t) = 0$ and is zero wherever $\tan(t)$ has a vertical asymptote.

  The graphs of the secant and cosecant functions are different from the cosine and sine functions' graphs in several ways, including that their range is the set of all real numbers y such that $y \geq 1$ and they have vertical asymptotes wherever the cosine and sine function, respectively, are zero.

- A trigonometric identity is an equation involving trigonometric functions that is true for every value of the variable for which the trigonometric functions are defined. For instance, $\tan^2(t) + 1 = \sec^2(t)$ for every real number t except $t = \frac{\pi}{2} \pm k\pi$. Identities offer us alternate perspectives on the same function. For instance, the function $f(t) = \sec^2(t) - \tan^2(t)$ is the same (at all points where f is defined) as the function whose value is always 1.

4.5.5 Exercises

1.   Find the exact value of each without using a calculator:

a) $\tan\left(\frac{7\pi}{4}\right)$

b) $\tan\left(\frac{5\pi}{4}\right)$

c) $\cot\left(\frac{5\pi}{3}\right)$

d) $\sec\left(\frac{3\pi}{4}\right)$

e) $\csc\left(\frac{\pi}{3}\right)$

2.   Suppose the angle θ is in the first quadrant, $0 \leq \theta \leq \pi/2$, and $\cos(\theta) = \frac{1}{8}$. Find exact values (as fractions, not decimal approximations) for the following.

(a) $\csc(\theta)$

(b) $\cot(\theta)$

3.   Suppose the angle θ is in the fourth quadrant, $\frac{3\pi}{2} \leq \theta \leq 2\pi$, and $\tan(\theta) = \frac{-2}{3}$. Find exact values (as fractions, not decimal approximations) for the following.

(a) $\sec(\theta)$

(b) $\sin(\theta)$

4. Let β be an angle in quadrant II that satisfies $\cos(\beta) = -\frac{12}{13}$. Determine the values of the other five trigonometric functions evaluated at β exactly and without evaluating any trigonometric function on a computational device.

   How do your answers change if β that lies in quadrant III?

5. For each of the following transformations of standard trigonometric functions, use your understanding of transformations to determine the domain, range, asymptotes, and period of the function, with careful justification. Then, check your results using *Desmos* or another graphing utility.

   a. $f(t) = 5\sec(t - \frac{\pi}{2}) + 3$

   b. $g(t) = -\frac{1}{3}\csc(2t) - 4$

   c. $h(t) = -7\tan(t + \frac{\pi}{4}) + 1$

   d. $j(t) = \frac{1}{2}\cot(4t) - 2$

6. In a right triangle with hypotenuse 1 and vertical leg x, with angle θ opposite x, determine the simplest expression you can for each of the following quantities in terms of x.

   a. $\sin(\theta)$

   d. $\tan(\theta)$

   b. $\sec(\theta)$

   e. $\cos(\arcsin(x))$

   c. $\csc(\theta)$

   f. $\cot(\arcsin(x))$

CHAPTER 5

Polynomial and Rational Functions

5.1 Infinity, limits, and power functions

Motivating Questions

- How can we use limit notation to succinctly express a function's behavior as the input increases without bound or as the function's value increases without bound?

- What are some important limits and trends involving ∞ that we can observe for familiar functions such as e^x, $\ln(x)$, x^2, and $\frac{1}{x}$?

- What is a power function and how does the value of the power determine the function's overall behavior?

In Section 3.2, we compared the behavior of the exponential functions $p(t) = 2^t$ and $q(t) = (\frac{1}{2})^t$, and observed in Figure 3.2.5 that as t increases without bound, $p(t)$ also increases without bound, while $q(t)$ approaches 0 (while having its value be always positive). We also introduced shorthand notation for describing these phenomena, writing

$$p(t) \to \infty \text{ as } t \to \infty$$

and

$$q(t) \to 0 \text{ as } t \to \infty.$$

It's important to remember that infinity is not itself a number. We use the "∞" symbol to represent a quantity that gets larger and larger with no bound on its growth.

We also know that the concept of infinity plays a key role in understanding the graphical behavior of functions. For instance, we've seen that for a function such as $F(t) = 72 - 45e^{-0.05t}$, $F(t) \to 72$ as $t \to \infty$, since $e^{-0.05t} \to 0$ as t increases without bound. The function F can be viewed as modeling the temperature of an object that is initially $F(0) = 72 - 45 = 27$ degrees that eventually warms to 72 degrees. The line $y = 72$ is thus a horizontal asymptote of the function F.

In Preview 5.1.1, we review some familiar functions and portions of their behavior that involve ∞.

Preview Activity 5.1.1. Complete each of the following statements with an appropriate number or the symbols ∞ or $-\infty$. Do your best to do so *without* using a graphing utility; instead use your understanding of the function's graph.

a. As $t \to \infty$, $e^{-t} \to$ _____.

b. As $t \to \infty$, $\ln(t) \to$ _____.

c. As $t \to \infty$, $e^t \to$ _____.

d. As $t \to 0^+$, $e^{-t} \to$ _____. (When we write $t \to 0^+$, this means that we are letting t get closer and closer to 0, but only allowing t to take on positive values.)

e. As $t \to \infty$, $35 + 53e^{-0.025t} \to$ _____.

f. As $t \to \frac{\pi}{2}^-$, $\tan(t) \to$ _____. (When we write $t \to \frac{\pi}{2}^-$, this means that we are letting t get closer and closer to $\frac{\pi}{2}^-$, but only allowing t to take on values that lie to the left of $\frac{\pi}{2}$.)

g. As $t \to \frac{\pi}{2}^+$, $\tan(t) \to$ _____. (When we write $t \to \frac{\pi}{2}^+$, this means that we are letting t get closer and closer to $\frac{\pi}{2}^+$, but only allowing t to take on values that lie to the right of $\frac{\pi}{2}$.)

5.1.1 Limit notation

When observing a pattern in the values of a function that correspond to letting the inputs get closer and closer to a fixed value or letting the inputs increase or decrease without bound, we are often interested in the behavior of the function "in the limit". In either case, we are considering an infinite collection of inputs that are themselves following a pattern, and we ask the question "how can we expect the function's output to behave if we continue?"

For instance, we have regularly observed that "as $t \to \infty$, $e^{-t} \to 0$," which means that by allowing t to get bigger and bigger without bound, we can make e^{-t} get as close to 0 as we'd like (without e^{-t} ever equalling 0, since e^{-t} is always positive).

Similarly, as seen in Figure 5.1.1 and Figure 5.1.2, we can make such observations as $e^t \to \infty$ as $t \to \infty$, $\ln(t) \to \infty$ as $t \to \infty$, and $\ln(t) \to -\infty$ as $t \to 0^+$. We introduce formal *limit notation* in order to be able to express these patterns even more succinctly.

Definition 5.1.3 Let L be a real number and f be a function. If we can make the value of $f(t)$ as close to L as we want by letting t increase without bound, we write

$$\lim_{t\to\infty} f(t) = L$$

and say that the **limit of f as t increases without bound is** L.

If the value of $f(t)$ increases without bound as t increases without bound, we instead write

$$\lim_{t\to\infty} f(t) = \infty.$$

Finally, if f doesn't increase without bound, doesn't decrease without bound, and doesn't

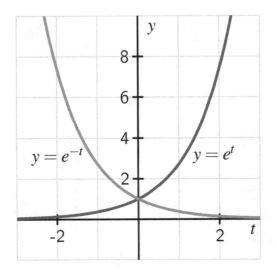

Figure 5.1.1: Plots of $y = e^t$ and $y = e^{-t}$.

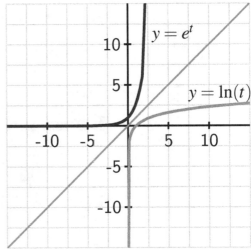

Figure 5.1.2: Plots of $y = e^t$ and $y = \ln(t)$.

approach a single value L as $t \to \infty$, we say that f **does not have a limit as** $t \to \infty$. ◊

We use limit notation in related, natural ways to express patterns we see in function behavior. For instance, we write $t \to -\infty$ when we let t decrease without bound, and $f(t) \to -\infty$ if f decreases without bound. We can also think about an input value t approaching a value a at which the function f is not defined. As one example, we write

$$\lim_{t \to 0^+} \ln(t) = -\infty$$

because the natural logarithm function decreases without bound as input values get closer and close to 0 (while always being positive), as seen in Figure 5.1.2.

In the situation where $\lim_{t \to \infty} f(t) = L$, this tells us that f has a horizontal asymptote at $y = L$ since the function's value approaches this fixed number as t increases without bound. Similarly, if we can say that $\lim_{t \to a} f(t) = \infty$, this shows that f has a vertical asymptote at $x = a$ since the function's value increases without bound as inputs approach the fixed number a.

For now, we are going to focus on the long-range behavior of certain basic, familiar functions and work to understand how they behave as the input increases or decreases without bound. Above we've used the input variable t in most of our previous work; going forward, we'll regularly use x as well.

> **Activity 5.1.2.** Complete the Table 5.1.4 by entering "∞," "−∞," "0," or "no limit" to identify how the function behaves as either x increases or decreases without bound. As much as possible, work to decide the behavior *without* using a graphing utility.

$f(x)$	$\lim_{x\to\infty} f(x)$	$\lim_{x\to-\infty} f(x)$
e^x		
e^{-x}		
$\ln(x)$		
x		
x^2		
x^3		
x^4		
$\frac{1}{x}$		
$\frac{1}{x^2}$		
$\sin(x)$		

Table 5.1.4: Some familiar functions and their limits as $x \to \infty$ or $x \to -\infty$.

5.1.2 Power functions

To date, we have worked with several families of functions: linear functions of form $y = mx + b$, quadratic functions in standard form, $y = ax^2 + bx + c$, the sinusoidal (trigono-metric) functions $y = a\sin(k(x-b)) + c$ or $y = a\cos(k(x-b)) + c$, transformed exponential functions such as $y = ae^{kx} + c$, and transformed logarithmic functions of form $y = a\ln(x) + c$. For trigonometric, exponential, and logarithmic functions, it was essential that we first un-derstood the behavior of the basic parent functions $\sin(x)$, $\cos(x)$, e^x, and $\ln(x)$. In order to build on our prior work with linear and quadratic functions, we now consider basic func-tions such as x, x^2, and additional powers of x.

Definition 5.1.5 A function of the form $f(x) = x^p$ where p is any real number is called a **power function**. ◊

We first focus on the case where p is a natural number (that is, a positive whole number).

Activity 5.1.3. Point your browser to the *Desmos* worksheet at http://gvsu.edu/s/ 0zu. In what follows, we explore the behavior of power functions of the form $y = x^n$ where $n \geq 1$.

a. Press the "play" button next to the slider labeled "n." Watch at least two loops of the animation and then discuss the trends that you observe. Write a careful sentence each for at least two different trends.

b. Click the icons next to each of the following 8 functions so that you can see all of $y = x$, $y = x^2$, ..., $y = x^8$ graphed at once. On the interval $0 < x < 1$, how do the graphs of x^a and x^b compare if $a < b$?

c. Uncheck the icons on each of the 8 functions to hide their graphs. Click the settings icon to change the domain settings for the axes, and change them to $-10 \le x \le 10$ and $-10,000 \le y \le 10,000$. Play the animation through twice and then discuss the trends that you observe. Write a careful sentence each for at least two different trends.

d. Click the icons next to each of the following 8 functions so that you can see all of $y = x$, $y = x^2$, ..., $y = x^8$ graphed at once. On the interval $x > 1$, how do the graphs of x^a and x^b compare if $a < b$?

In the situation where the power p is a negative integer (i.e., a negative whole number), power functions behave very differently. This is because of the property of exponents that states

$$x^{-n} = \frac{1}{x^n}$$

so for a power function such as $p(x) = x^{-2}$, we can equivalently consider $p(x) = \frac{1}{x^2}$. Note well that for these functions, their domain is the set of all real numbers except $x = 0$. Like with power functions with positive whole number powers, we want to know how power functions with negative whole number powers behave as x increases without bound, as well as how the functions behave near $x = 0$.

Activity 5.1.4. Point your browser to the *Desmos* worksheet at http://gvsu.edu/s/0zv. In what follows, we explore the behavior of power functions $y = x^n$ where $n \le -1$.

a. Press the "play" button next to the slider labeled "n." Watch two loops of the animation and then discuss the trends that you observe. Write a careful sentence each for at least two different trends.

b. Click the icons next to each of the following 8 functions so that you can see all of $y = x^{-1}$, $y = x^{-2}$, ..., $y = x^{-8}$ graphed at once. On the interval $1 < x$, how do the functions x^a and x^b compare if $a < b$? (Be careful with negative numbers here: e.g., $-3 < -2$.)

c. How do your answers change on the interval $0 < x < 1$?

d. Uncheck the icons on each of the 8 functions to hide their graphs. Click the settings icon to change the domain settings for the axes, and change them to $-10 \le x \le 10$ and $-10,000 \le y \le 10,000$. Play the animation through twice and then discuss the trends that you observe. Write a careful sentence each for at least two different trends.

e. Explain why $\lim_{x \to \infty} \frac{1}{x^n} = 0$ for any choice of $n = 1, 2, \ldots$.

5.1.3 Summary

- The notation

$$\lim_{x \to \infty} f(x) = L$$

means that we can make the value of $f(x)$ as close to L as we'd like by letting x be sufficiently large. This indicates that the value of f eventually stops changing much and tends to a single value, and thus $y = L$ is a horizontal asymptote of the function f.

Similarly, the notation

$$\lim_{x \to a} f(x) = \infty$$

means that we can make the value of $f(x)$ as large as we'd like by letting x be sufficiently close, but not equal, to a. This unbounded behavior of f near a finite value a indicates that f has a vertical asymptote at $x = a$.

- We summarize some key behavior of familiar basic functions with limits as x increases without bound in Table 5.1.6.

$f(x)$	$\lim_{x \to \infty} f(x)$	$\lim_{x \to -\infty} f(x)$
e^x	∞	0
e^{-x}	0	∞
$\ln(x)$	∞	NA[1]
x	∞	$-\infty$
x^2	∞	∞
x^3	∞	$-\infty$
x^4	∞	∞
$\frac{1}{x}$	0	0
$\frac{1}{x^2}$	0	0
$\sin(x)$	no limit[2]	no limit

Table 5.1.6: Some familiar functions and their limits as $x \to \infty$ or $x \to -\infty$.

Additionally, Table 5.1.7 summarizes some key familiar function behavior where the function's output increases or decreases without bound as x approaches a fixed number not in the function's domain.

---

[1]Because the domain of the natural logarithm function is only positive real numbers, it doesn't make sense to even consider this limit.

[2]Because the sine function neither increases without bound nor approaches a single value, but rather keeps oscillating through every value between −1 and 1 repeatedly, the sine function does not have a limit as $x \to \infty$.

$f(x)$	$\lim_{x \to a^-} f(x)$	$\lim_{x \to a^+} f(x)$
$\ln(x)$	NA	$\lim_{x \to 0^+} \ln(x) = -\infty$
$\frac{1}{x}$	$\lim_{x \to 0^-} \frac{1}{x} = -\infty$	$\lim_{x \to 0^+} \frac{1}{x} = \infty$
$\frac{1}{x^2}$	$\lim_{x \to 0^-} \frac{1}{x^2} = \infty$	$\lim_{x \to 0^+} \frac{1}{x^2} = \infty$
$\tan(x)$	$\lim_{x \to \frac{\pi}{2}^-} \tan(x) = \infty$	$\lim_{x \to \frac{\pi}{2}^+} \tan(x) = -\infty$
$\sec(x)$	$\lim_{x \to \frac{\pi}{2}^-} \sec(x) = \infty$	$\lim_{x \to \frac{\pi}{2}^+} \sec(x) = -\infty$
$\csc(x)$	$\lim_{x \to 0^-} \sec(x) = -\infty$	$\lim_{x \to 0^+} \sec(x) = \infty$

Table 5.1.7: Some familiar functions and their limits involving ∞ as $x \to a$ where a is not in the function's domain.

- A power function is a function of the form $f(x) = x^p$ where p is any real number. For the two cases where p is a positive whole number or a negative whole number, it is straightforward to summarize key trends in power functions' behavior.

  ○ If $p = 1, 2, 3, \ldots$, then the domain of $f(x) = x^p$ is the set of all real numbers, and as $x \to \infty$, $f(x) \to \infty$. For the limit as $x \to -\infty$, it matters whether p is even or odd: if p is even, $f(x) \to \infty$ as $x \to -\infty$; if p is odd, $f(x) \to -\infty$ as $x \to \infty$. Informally, all power functions of form $f(x) = x^p$ where p is a positive even number are "U-shaped", while all power functions of form $f(x) = x^p$ where p is a positive odd number are "chair-shaped".

  ○ If $p = -1, -2, -3, \ldots$, then the domain of $f(x) = x^p$ is the set of all real numbers *except* $x = 0$, and as $x \to \pm\infty$, $f(x) \to 0$. This means that each such power function with a negative whole number exponent has a horizontal asymptote of $y = 0$. Regardless of the value of p ($p = -1, -2, -3, \ldots$), $\lim_{x \to 0^+} f(x) = \infty$. But when we approach 0 from the negative side, it matters whether p is even or odd: if p is even, $f(x) \to \infty$ as $x \to 0^-$; if p is odd, $f(x) \to -\infty$ as $x \to 0^-$. Informally, all power functions of form $f(x) = x^p$ where p is a negative odd number look similar to $\frac{1}{x}$, while all power functions of form $f(x) = x^p$ where p is a negative even number look similar to $\frac{1}{x^2}$.

5.1.4 Exercises

1. Find the long run behavior of each of the following functions.

   (a) As $x \to -\infty$, $-6x^{-3} \to$ _____

   (b) As $x \to \infty$, $(14 - 6x^3) \to$ _____

2. Find:

   (a) $\lim_{t \to -\infty} \left(\frac{1}{t^2} + 3\right)$

   (b) $\lim_{t \to \infty} 3\frac{1}{y}$

3. Is the function $g(x) = \dfrac{(-x^3)^5}{9}$ a power function? If it is, write it in the form $g(x) = kx^p$.

4. We've observed that several different familiar functions grow without bound as $x \to \infty$, including $f(x) = \ln(x)$, $g(x) = x^2$, and $h(x) = e^x$. In this exercise, we compare and contrast how these three functions grow.

    a. Use a computational device to compute decimal expressions for $f(10)$, $g(10)$, and $h(10)$, as well as $f(100)$, $g(100)$, and $h(100)$. What do you observe?

    b. For each of f, g, and h, how large an input is needed in order to ensure that the function's output value is at least 10^{10}? What do these values tell us about how each function grows?

    c. Consider the new function $r(x) = \dfrac{g(x)}{h(x)} = \dfrac{x^2}{e^x}$. Compute $r(10)$, $r(100)$, and $r(1000)$. What do the results suggest about the long-range behavior of r? What is surprising about this, in light of the fact that both x^2 and e^x grow without bound?

5. Consider the familiar graph of $f(x) = \frac{1}{x}$, which has a vertical asypmtote at $x = 0$ and a horizontal asymptote at $y = 0$, as pictured in Figure 5.1.8. In addition, consider the similarly-shaped function g shown in Figure 5.1.9, which has vertical asymptote $x = -1$ and horizontal asymptote $y = -2$.

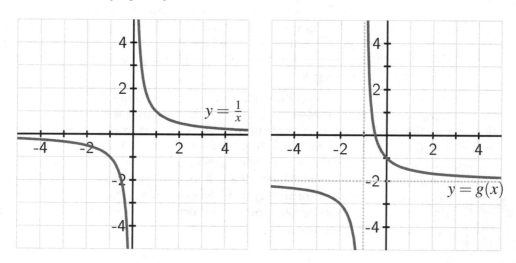

Figure 5.1.8: A plot of $y = f(x) = \frac{1}{x}$.   **Figure 5.1.9:** A plot of a related function $y = g(x)$.

    a. How can we view g as a transformation of f? Explain, and state how g can be expressed algebraically in terms of f.

    b. Find a formula for g as a function of x. What is the domain of g?

    c. Explain algebraically (using the form of g from (b)) why $\lim_{x \to \infty} g(x) = -2$ and $\lim_{x \to -1^+} g(x) = \infty$.

    d. What if a function h (again of a similar shape as f) has vertical asymptote $x = 5$ and horizontal asymtote $y = 10$. What is a possible formula for $h(x)$?

e. Suppose that $r(x) = \frac{1}{x+35} - 27$. *Without* using a graphing utility, how do you expect the graph of r to appear? Does it have a horizontal asymptote? A vertical asymptote? What is its domain?

6. Power functions can have powers that are not whole numbers. For instance, we can consider such functions as $f(x) = x^{2.4}$, $g(x) = x^{2.5}$, and $h(x) = x^{2.6}$.

   a. Compare and contrast the graphs of f, g, and h. How are they similar? How are they different? (There is a lot you can discuss here.)

   b. Observe that we can think of $f(x) = x^{2.4}$ as $f(x) = x^{24/10} = x^{12/5}$. In addition, recall by exponent rules that we can also view f as having the form $f(x) = \sqrt[5]{x^{12}}$. Write g and h in similar forms, and explain why g has a different domain than f and h.

   c. How do the graphs of f, g, and h compare to the graphs of $y = x^2$ and $y = x^3$? Why are these natural functions to use for comparison?

   d. Explore similar questions for the graphs of $p(x) = x^{-2.4}$, $q(x) = x^{-2.5}$, and $r(x) = x^{-2.6}$.

5.2 Polynomials

Motivating Questions

- What properties of a polynomial function can we deduce from its algebraic structure?

- What is a sign chart and how does it help us understand a polynomial function's behavior?

- How do zeros of multiplicity other than 1 impact the graph of a polynomial function?

We know that linear functions are the simplest of all functions we can consider: their graphs have the simplest shape, their average rate of change is always constant (regardless of the interval chosen), and their formula is elementary. Moreover, computing the value of a linear function only requires multiplication and addition.

If we think of a linear function as having formula $L(x) = b + mx$, and the next-simplest functions, quadratic functions, as having form $Q(x) = c + bx + ax^2$, we can see immediate parallels between their respective forms and realize that it's natural to consider slightly more complicated functions by adding additional power functions.

Indeed, if we instead view linear functions as having form

$$L(x) = a_0 + a_1 x$$

(for some constants a_0 and a_1) and quadratic functions as having form

$$Q(x) = a_0 + a_1 x + a_2 x^2$$

(for some constants a_0, a_1, and a_2), then it's natural to think about more general functions of this same form, but with additional power functions included.

Definition 5.2.1 Given real numbers $a_0, a_1, \ldots, a_n$ where $a_n \neq 0$, we say that the function

$$P(x) = a_0 + a_1 x + a_2 x^2 + \cdots + a_{n-1} x^{n-1} + a_n x^n$$

is a **polynomial of degree** n. In addition, we say that the values of a_i are the **coefficients** of the polynomial and the individual power functions $a_i x^i$ are the **terms** of the polynomial. Any value of x for which $P(x) = 0$ is called a **zero** of the polynomial. ◊

Example 5.2.2 The polyomial function $P(x) = 3 - 7x + 4x^2 - 2x^3 + 9x^5$ has degree 5, its constant term is 3, and its linear term is $-7x$. □

Since a polynomial is simply a sum of constant multiples of various power functions with positive integer powers, we often refer to those individual terms by referring to their individual degrees: the linear term, the quadratic term, and so on. In addition, since the domain of any power function of the form $p(x) = x^n$ where n is a positive whole number is the set of all real numbers, it's also true the the domain of any polynomial function is the set of all real numbers.

Preview Activity 5.2.1. Point your browser to the *Desmos* worksheet at http://gvsu.edu/s/0zy. There you'll find a degree 4 polynomial of the form $p(x) = a_0 + a_1x + a_2x^2 + a_3x^3 + a_4x^4$, where $a_0, \ldots, a_4$ are set up as sliders. In the questions that follow, you'll experiment with different values of $a_0, \ldots, a_4$ to investigate different possible behaviors in a degree 4 polynomial.

a. What is the largest number of distinct points at which $p(x)$ can cross the x-axis?

   For a polynomial p, we call any value r such that $p(r) = 0$ a **zero** of the polynomial. Report the values of $a_0, \ldots, a_4$ that lead to that largest number of zeros for $p(x)$.

b. What other numbers of zeros are possible for $p(x)$? Said differently, can you get each possible number of fewer zeros than the largest number that you found in (a)? Why or why not?

c. We say that a function has a **turning point** if the function changes from decreasing to increasing or increasing to decreasing at the point. For example, any quadratic function has a turning point at its vertex.

   What is the largest number of turning points that $p(x)$ (the function in the *Desmos* worksheet) can have? Experiment with the sliders, and report values of $a_0, \ldots, a_4$ that lead to that largest number of turning points for $p(x)$.

d. What other numbers of turning points are possible for $p(x)$? Can it have no turning points? Just one? Exactly two? Experiment and explain.

e. What long-range behavior is possible for $p(x)$? Said differently, what are the possible results for $\lim_{x \to -\infty} p(x)$ and $\lim_{x \to \infty} p(x)$?

f. What happens when we plot $y = a_4x^4$ in and compare $p(x)$ and a_4x^4? How do they look when we zoom out? (Experiment with different values of each of the sliders, too.)

5.2.1 Key results about polynomial functions

Our observations in Preview Activity 5.2.1 generalize to polynomials of any degree. In particular, it is possible to prove the following general conclusions regarding the number of zeros, the long-range behavior, and the number of turning points any polynomial of degree n.

> **The Fundamental Theorem of Algebra.**
>
> For any degree n polynomial $p(x) = a_0 + a_1x + \cdots + a_{n-1}x^{n-1} + a_nx^n$, has at most n real zeros.[1]

---

[1]We can actually say even more: if we allow the zeros to be complex numbers, then every degree n polynomial has *exactly* n zeros, provided we count zeros according to their multiplicity. For example, the polynomial $p(x) = (x - 1)^2 = x^2 - 2x + 1$ because it has a zero of multiplicity two at $x = 1$.

We know that each of the power functions $x, x^2, \ldots, x^n$ grow without bound as $x \to \infty$. Intuitively, we sense that x^5 grows faster than x^4 (and likewise for any comparison of a higher power to a lower one). This means that for large values of x, the most important term in any polynomial is its highest order term, as we saw in Preview Activity 5.2.1 when we compared $p(x) = a_0 + a_1 x + a_2 x^2 + a_3 x^3 + a_4 x^4$ and $y = a_4 x^4$.

The long-range behavior of a polynomial.

For any degree n polynomial $p(x) = a_0 + a_1 x + \cdots + a_{n-1} x^{n-1} + a_n x^n$, its long-range behavior is the same as its highest-order term $q(x) = a_n x^n$. Thus, any polynomial of even degree appears "U-shaped" ($\cup$ or $\cap$, like x^2 or $-x^2$) when we zoom way out, and any polynomial of odd degree appears "chair-shaped" (like x^3 or $-x^3$) when we zoom way out.

In Figure 5.2.4, we see how the degree 7 polynomial pictured there (and in Figure 5.2.3 as well) appears to look like $q(x) = -x^7$ as we zoom out.

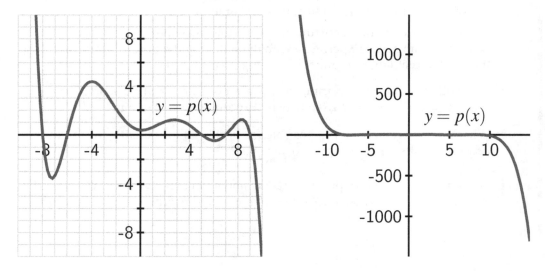

Figure 5.2.3: Plot of a degree 7 polynomial function p.

Figure 5.2.4: Plot of the same degree 7 polynomial function p, but zoomed out.

Finally, a key idea from calculus justifies the fact that the maximum number of turning points of a degree n polynomial is $n - 1$, as we conjectured in the degree 4 case in Preview Activity 5.2.1. Moreover, the only possible numbers of turning points must have the same parity as $n-1$; that is, if $n-1$ is even, then the number of turning points must be even, and if instead $n - 1$ is odd, the number of turning points must also be odd. For instance, for the degree 7 polynomial in Figure 5.2.3, we know that it is chair-shaped, with one end up and one end down. There could be zero turning points and the function could always decrease. But if there is at least one, then there must be a second, since if there were only one the function would decrease and then increase without turning back, which would force the graph to appear U-shaped.

The turning points of a polynomial.

For any degree n polynomial $p(x) = a_0 + a_1x + \cdots + a_{n-1}x^{n-1} + a_nx^n$, if n is even, its number of turning points is exactly one of $n - 1$, $n - 3$, ..., 1, and if n is odd, its number of turning points is exactly one of $n - 1$, $n - 3$, ..., 0.

Activity 5.2.2. By experimenting with coefficients in *Desmos*, find a formula for a polynomial function that has the stated properties, or explain why no such polynomial exists. (If you enter p(x)=a+bx+cx^2+dx^3+fx^4+gx^5 in *Desmos*[2], you'll get prompted to add sliders that make it easy to explore a degree 5 polynomial.)

a. A polynomial p of degree 5 with exactly 3 real zeros, 4 turning points, and such that $\lim_{x \to -\infty} p(x) = +\infty$ and $\lim_{x \to \infty} p(x) = -\infty$.

b. A polynomial p of degree 4 with exactly 4 real zeros, 3 turning points, and such that $\lim_{x \to -\infty} p(x) = +\infty$ and $\lim_{x \to \infty} p(x) = -\infty$.

c. A polynomial p of degree 6 with exactly 2 real zeros, 3 turning points, and such that $\lim_{x \to -\infty} p(x) = -\infty$ and $\lim_{x \to \infty} p(x) = -\infty$.

d. A polynomial p of degree 5 with exactly 5 real zeros, 3 turning points, and such that $\lim_{x \to -\infty} p(x) = +\infty$ and $\lim_{x \to \infty} p(x) = -\infty$.

5.2.2 Using zeros and signs to understand polynomial behavior

Just like a quadratic function can be written in different forms (standard: $q(x) = ax^2 + bx + c$, vertex: $q(x) = a(x - h)^2 + k$, and factored: $q(x) = a(x - r_1)(x - r_2)$), it's possible to write a polynomial function in different forms and to gain information about its behavior from those different forms. In particular, if we know all of the zeros of a polynomial function, we can write its formula in factored form, which gives us a deeper understanding of its graph.

The Zero Product Property states that if two or more numbers are multiplied together and the result is 0, then at least one of the numbers must be 0. We use the Zero Product Property regularly with polynomial functions. If we can determine all n zeros of a degree n polynomial, and we call those zeros $r_1, r_2, \ldots, r_n$, we can write

$$p(x) = a_n(x - r_1)(x - r_2) \cdots (x - r_2).$$

Moreover, if we are given a polynomial in this factored form, we can quickly determine its zeros. For instance, if $p(x) = 2(x + 7)(x + 1)(x - 2)(x - 5)$, we know that the only way $p(x) = 0$ is if at least one of the factors $(x + 7)$, $(x + 1)$, $(x - 2)$, or $(x - 5)$ equals 0, which implies that $x = -7$, $x = -1$, $x = 2$, or $x = 5$. Hence, from the factored form of a polynomial, it is straightforward to identify the polynomial's zeros, the x-values at which its graph crosses the x-axis. We can also use the factored form of a polynomial to develop what we call a *sign chart*, which we demonstrate in Example 5.2.5.

---

[2]We skip using e as one of the constants since *Desmos* reserves e as the Euler constant.

Example 5.2.5 Consider the polynomial function $p(x) = k(x - 1)(x - a)(x - b)$. Suppose we know that $1 < a < b$ and that $k < 0$. Fully describe the graph of p without the aid of a graphing utility.

Solution. Since $p(x) = k(x - 1)(x - a)(x - b)$, we immediately know that p is a degree 3 polynomial with 3 real zeros: $x = 1, a, b$. We are given that $1 < a < b$ and in addition that $k < 0$. If we expand the factored form of $p(x)$, it has form $p(x) = kx^3 + \cdots$, and since we know that when we zoom out, $p(x)$ behaves like kx^3, we know that with $k < 0$ it follows $\lim_{x \to -\infty} p(x) = +\infty$ and $\lim_{x \to \infty} p(x) = -\infty$.

Since p is degree 3 and we know it has zeros at $x = 1, a, b$, we know there are no other locations where $p(x) = 0$. Thus, on any interval between two zeros (or to the left of the least or the right of the greatest), the polynomial cannot change sign. We now investigate, interval by interval, the sign of the function.

When $x < 1$, it follows that $x - 1 < 0$. In addition, since $1 < a < b$, when $x < 1$, x lies to the left of 1, a, and b, which also makes $x - a$ and $x - b$ negative. Moreover, we know that the constant $k < 0$. Hence, on the interval $x < 1$, all four terms in $p(x) = k(x-1)(x-a)(x-b)$ are negative, which we indicate by writing "$- - --$" in that location on the sign chart pictured in Figure 5.2.6.

In addition, since there are an even number of negative terms in the product, the overall product's sign is positive, which we indicate by the single "$+$" beneath "$- - --$", and by writing "POS" below the coordinate axis.

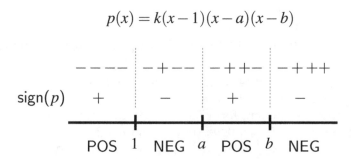

$$p(x) = k(x-1)(x-a)(x-b)$$

Figure 5.2.6: A sign chart for the polynomial function $p(x) = k(x - 1)(x - a)(x - b)$.

We now proceed to the other intervals created by the zeros. On $1 < x < a$, the term $(x - 1)$ has become positive, since $x > 1$. But both $x - a$ and $x - b$ are negative, as is the constant k, and thus we write "$- + --$" for this interval, which has overall sign "$-$", as noted in the figure. Similar reasoning completes the diagram.

From all of the information we have deduced about p, we conclude that regardless of the locations of a and b, the graph of p must look like the curve shown in Figure 5.2.7.

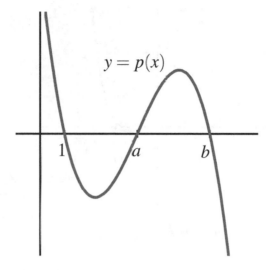

Figure 5.2.7: The graph of the polynomial function $p(x) = k(x - 1)(x - a)(x - b)$.

□

Activity 5.2.3. Consider the polynomial function given by

$$p(x) = 4692(x + 1520)(x^2 + 10000)(x - 3471)^2(x - 9738).$$

a. What is the degree of p? How can you tell *without* fully expanding the factored form of the function?

b. What can you say about the sign of the factor $(x^2 + 10000)$?

c. What are the zeros of the polynomial p?

d. Construct a sign chart for p by using the zeros you identified in (b) and then analyzing the sign of each factor of p.

e. Without using a graphing utility, construct an approximate graph of p that has the zeros of p carefully labeled on the x-axis.

f. Use a graphing utility to check your earlier work. What is challenging or misleading when using technology to graph p?

5.2.3 Multiplicity of polynomial zeros

In Activity 5.2.3, we found that one of the zeros of the polynomial $p(x) = 4692(x + 1520)(x^2 + 10000)(x - 3471)^2(x - 9738)$ leads to different behavior of the function near that zero than we've seen in other situations. We now consider the more general situation where a polynomial has a repeated factor of the form $(x - r)^n$. When $(x - r)^n$ is a factor of a polynomial

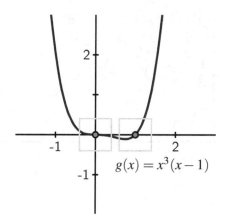

Figure 5.2.8: A plot of $g(x) = x^3(x - 1)$ with zero $x = 0$ of multiplicity 3 and $x = 1$ of multiplicity 1.

p, we say that p has a **zero of multiplicity** n at $x = r$.

To see the impact of repeated factors, we examine a collection of degree 4 polynomials that each have 4 real zeros. We start with the simplest of all, the function $f(x) = x^4$, whose zeros are $x = 0, 0, 0, 0$. Because the factor "$x-0$" is repeated 4 times, the zero $x = 0$ has multiplicity 4.

Next we consider the degree 4 polynomial $g(x) = x^3(x - 1)$, which has a zero of multiplicity 3 at $x = 0$ and a zero of multiplicity 1 at $x = 1$. Observe that in Figure 5.2.9, the up-close plot near the zero $x = 0$ of multiplicity 3, the polynomial function g looks similar to the basic cubic polynomial $-x^3$. In addition, in Figure 5.2.10, we observe that if we zoom in even futher on the zero of multiplicity 1, the function g looks roughly linear, like a degree 1 polynomial. This type of behavior near repeated zeros turns out to hold in other cases as well.

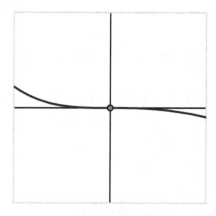 

Figure 5.2.9: A plot of $g(x) = x^3(x - 1)$ zoomed in on the zero $x = 0$ of multiplicity 3.

Figure 5.2.10: A plot of $g(x) = x^3(x - 1)$ zoomed in on the zero $x = 1$ of multiplicity 1.

If we next let $h(x) = x^2(x-1)^2$, we see that h has two distinct real zeros, each of multiplicity 2. The graph of h in Figure 5.2.11 shows that h behaves similar to a basic quadratic function near each of those zeros and thus shows U-shaped behavior nearby. If instead we let $k(x) = x^2(x-1)(x+1)$, we see approximately linear behavior near $x = -1$ and $x = 1$ (the zeros of multiplicity 1), and quadratic (U-shaped) behavior near $x = 0$ (the zero of multiplicity 2), as seen in Figure 5.2.12.

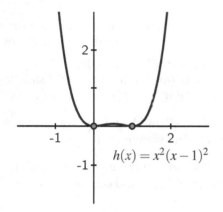

Figure 5.2.11: Plot of $h(x) = x^2(x-1)^2$ with zeros $x = 0$ and $x = 1$ of multiplicity 2.

Figure 5.2.12: Plot of $k(x) = x^2(x-1)(x+1)$ with zeros $x = 0$ of multiplicity 2 and $x = -1$ and $x = 1$ of multiplicity 1.

Finally, if we consider $m(x) = (x+1)x(x-1)(x-2)$, which has 4 distinct real zeros each of multiplicity 1, we observe in Figure 5.2.13 that zooming in on each zero individually, the function demonstrates approximately linear behavior as it passes through the x-axis.

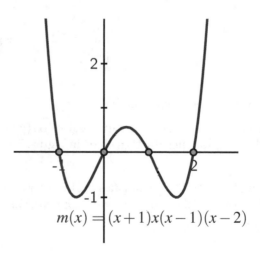

$$m(x) = (x+1)x(x-1)(x-2)$$

Figure 5.2.13: Plot of $m(x) = (x+1)x(x-1)(x-2)$ with 4 distinct zeros of multiplicity 1.

Our observations with polynomials of degree 4 in the various figures above generalize to polynomials of any degree.

> **Polynomial zeros of multiplicity n.**
>
> If $(x - r)^n$ is a factor of a polynomial p, then $x = r$ is a zero of p of multiplicity n, and near $x = r$ the graph of p looks like either $-x^n$ or x^n does near $x = 0$. That is, the shape of the graph near the zero is determined by the multiplicity of the zero.

Activity 5.2.4. For each of the following prompts, try to determine a formula for a polynomial that satisfies the given criteria. If no such polynomial exists, explain why.

a. A polynomial f of degree 10 whose zeros are $x = -12$ (multiplicity 3), $x = -9$ (multiplicity 2), $x = 4$ (multiplicity 4), and $x = 10$ (multiplicity 1), and f satisfies $f(0) = 21$. What can you say about the values of $\lim_{x \to -\infty} f(x)$ and $\lim_{x \to \infty} f(x)$?

b. A polynomial p of degree 9 that satisfies $p(0) = -2$ and has the graph shown in Figure 5.2.14. Assume that all of the zeros of p are shown in the figure.

c. A polynomial q of degree 8 with 3 distinct real zeros (possibly of different multiplicities) such that q has the sign chart in Figure 5.2.15 and satisfies $q(0) = -10$.

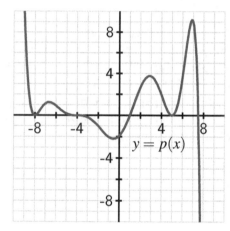

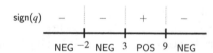

Figure 5.2.15: A sign chart for the polynomial q.

Figure 5.2.14: A degree 9 polynomial p.

d. A polynomial q of degree 9 with 3 distinct real zeros (possibly of different multiplicities) such that q satisfies the sign chart in Figure 5.2.15 and satisfies $q(0) = -10$.

e. A polynomial p of degree 11 that satisfies $p(0) = -2$ and p has the graph shown in Figure 5.2.14. Assume that all of the zeros of p are shown in the figure.

5.2.4 Summary

- From a polynomial function's algebraic structure, we can deduce several key traits of the function.

○ If the function is in standard form, say

$$p(x) = a_0 + a_1x + a_2x^2 + \cdots + a_{n-1}x^{n-1} + a_nx^n,$$

we know that its degree is n and that when we zoom out, p looks like a_nx^n and thus has the same long-range behavior as a_nx^n. Thus, p is chair-shaped if n is odd and U-shaped if n is even. Whether $\lim_{n \to \infty} p(x)$ is $+\infty$ or $-\infty$ depends on the sign of a_n.

○ If the function is in factored form, say

$$p(x) = a_n(x - r_1)(x - r_2) \cdots (x - r_n)$$

(where the r_i's are possibly not distinct and possibly complex), we can quickly determine both the degree of the polynomial (n) and the locations of its zeros, as well as their multiplicities.

• A sign chart is a visual way to identify all of the locations where a function is zero along with the sign of the function on the various intervals the zeros create. A sign chart gives us an overall sense of the graph of the function, but without concerning ourselves with any specific values of the function besides the zeros. For a sample sign chart, see Figure 5.2.6.

• When a polynomial p has a repeated factor such as

$$p(x) = (x - 5)(x - 5)(x - 5) = (x - 5)^3,$$

we say that $x = 5$ is a zero of multiplicity 3. At the point $x = 5$ where p will cross the x-axis, up close it will look like a cubic polynomial and thus be chair-shaped. In general, if $(x - r)^n$ is a factor of a polynomial p so that $x = r$ is a zero of multiplicity n, the polynomial will behave near $x = r$ like the polynomial x^n behaves near $x = 0$.

5.2.5 Exercises

1. Are the functions below polynomials? If they are, find their degree.

   $f(x) = 6^x + 4$

   $g(x) = x^6 + 4$

2. Are the functions below polynomials? If they are, find their degree.

   $f(x) = 6x^{3.7} + 2$

   $g(x) = 6x^2 + 3.7$

   $h(x) = 2x^{-6} + 3.7$

3. Let $y = 9x^6 - 3961x^2 + 6$.

   (a) What power function does the function above resemble?

   (b) Describe the long-run behavior of the polynomial.

   y goes to _____ as $x \to \infty$

y goes to _____ as $x \to -\infty$.

4. Let $y = 8x^3 + \frac{6x^4}{x-6} - 7x^5 + 1$.

    (a) What power function does the function above resemble?

    (b) Describe the long-run behavior of the polynomial.

    y goes to _____ as $x \to \infty$.

    y goes to _____ as $x \to -\infty$.

5. Estimate the zero(s) of $f(x) = x^4 + 16x^3 + 93x^2 + 231x + 206$.

6. Suppose $f(x) = x^2(7 - 8x^9)$.

    (a) Find roots of $f(x)$.

    (b) As $x \to \infty$, $f(x) \to$ _____

    (b) As $x \to -\infty$, $f(x) \to$ _____

7. Suppose $f(x) = (5 - 6x)(2x - 4)^2$.

    (a) Find the roots of $f(x)$.

    (b) As $x \to \infty$, $f(x) \to$ _____

    (b) As $x \to -\infty$, $f(x) \to$ _____

8. Consider the polynomial function given by

    $$p(x) = 0.0005(x + 21.7)^3(x - 20.9)^2(x - 31.4)(x^2 + 100).$$

    (a) What is the degree of p?

    (b) What are the real zeros of p? State them with multiplicity.

    (c) Construct a carefully labeled sign chart for $p(x)$.

    (d) Plot the function p in *Desmos*. Are the zeros obvious from the graph? How do you have to adjust the window in order to tell? Even in an adjusted window, can you tell them exactly from the graph?

    (e) Now consider the related but different polynomial

    $$q(x) = -0.0005(x + 21.7)^3(x - 20.9)^2(x - 31.4)(x^2 + 100)(x - 92.3).$$

    What is the degree of q? What are the zeros of q? What is obvious from its graph and what is not?

9. Consider the (non-polynomial) function $r(x) = e^{-x^2}(x^2 + 1)(x - 2)(x - 3)$.

    (a) What are the zeros of $r(x)$? (Hint: is $e^{\square}$ ever equal to zero?)

    (b) Construct a sign chart for $r(x)$.

    (c) Plot $r(x)$ in *Desmos*. Is the sign and overall behavior of r obvious from the plot? Why or why not?

(d) From the graph, what appears to be the value of $\lim_{x \to \infty} r(x)$? Why is this surprising in light of the behavior of $f(x) = (x^2 + 1)(x - 2)(x - 3)$ as $x \to \infty$?

10. In each following question, find a formula for a polynomial with certain properties, generate a plot that demonstrates you've found a function with the given specifications, and write several sentences to explain your thinking.

   (a) A quadratic function q has zeros at $x = 7$ and $x = 11$ and its y-value at its vertex is 42.

   (b) A polynomial r of degree 4 has zeros at $x = 3$ and $x = 5$, both of multiplicity 2, and the function has a y-intercept at the point $(0, 28)$.

   (c) A polynomial f has degree 11 and the following zeros: zeros of multiplicity 1 at $x = 3$ and $x = 5$, zeros of multiplicity 2 at $x = 2$ and $x = 3$, and a zero of multiplicity 3 at $x = 1$. In addition, $\lim_{x \to \infty} f(x) = -\infty$.

   (d) A polynomial g has its graph given in Figure 5.2.16 below. Determine a possible formula for $g(x)$ where the polynomial you find has the lowest possible degree to match the graph. What is the degree of the function you find?

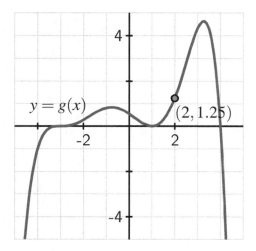

Figure 5.2.16: A polynomial function g.

11. Like we have worked to understand families of functions that involve parameters such as $p(t) = a \cos(k(t - b)) + c$ and $F(t) = a + be^{-kt}$, we are often interested in polynomials that involve one or more parameters and understanding how those parameters affect the function's behavior.

   For example, let $a > 0$ be a positive constant, and consider $p(x) = x^3 - a^2 x$.

   (a) What is the degree of p?

   (b) What is the long-term behavior of p? State your responses using limit notation.

   (c) In terms of the constant a, what are the zeros of p?

(d) Construct a carefully labeled sign chart for p.

(e) How does changing the value of a affect the graph of p?

5.3 Modeling with polynomial functions

Motivating Questions

- Why do polynomials arise naturally in the study of problems involving the volume and surface area of three-dimensional containers such as boxes and cylinders?

- How can polynomial functions be used to approximate non-polynomial curves and functions?

Polynomial functions are the simplest of all functions in mathematics in part because they only involve multiplication and addition. In any applied setting where we can formulate key ideas using only those arithmetic operations, it's natural that polynomial functions model the corresponding phenomena. For example, in Activity 1.2.2, we saw that for a spherical tank of radius 4 m filling with water, the volume of water in the tank at a given instant, V, is a function of the depth, h, of the water in the tank at the same moment according to the formula

$$V = f(h) = \frac{\pi}{3}h^2(12 - h).$$

The function f is a polynomial of degree 3 with a repeated zero at $h = 0$ and an additional zero at $h = 12$. Because the tank has a radius of 4, its total height is h, and thus the model $V = f(h) = \frac{\pi}{3}h^2(12 - h)$ is only valid on the domain $0 \le h \le 8$. This polynomial function tells us how the volume of water in the tank changes as h changes.

In other similar situations where we consider the volume of a box, tank, or other three-dimensional container, polynomial functions frequently arise. To develop a model function that represents a physical situation, we almost always begin by drawing one or more diagrams of the situation and then introduce one or more variables to represent quantities that are changing. From there, we explore relationships that are present and work to express one of the quantities in terms of the other(s).

> **Preview Activity 5.3.1.** A piece of cardboard that is 12×18 (each measured in inches) is being made into a box without a top. To do so, squares are cut from each corner of the cardboard and the remaining sides are folded up.
>
> a. Let x be the side length of the squares being cut from the corners of the cardboard. Draw a labeled diagram that shows the given information and the variable being used.
>
> b. Determine a formula for the function V whose output is the volume of the box that results from a square of size $x \times x$ being cut from each corner of the cardboard.
>
> c. What familiar kind of function is V?
>
> d. What is the largest size of a square that could be cut from the cardboard and still have a resulting box?
>
> e. What are the zeros of V? What is the domain of the model V in the context of

the rectangular box?

5.3.1 Volume, surface area, and constraints

In Preview Activity 5.3.1, we worked with a rectangular box being built by folding cardboard. One of the key principles we needed to use was the fact that the volume of a rectangular box of length l, width w, and height h is

$$V = lwh. \tag{5.3.1}$$

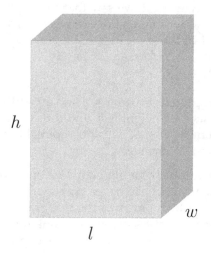

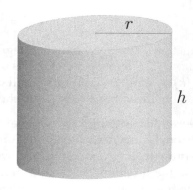

Figure 5.3.1: A rectangular box.   **Figure 5.3.2:** A circular cylinder.

One way to remember the formula for the area of a rectangular box is "area of the base times the height". This principle extends to other three-dimensional shapes that have constant cross-sectional area. For instance, the volume of a circular cylinder with radius r and height h is

$$V = \pi r^2 h \tag{5.3.2}$$

since the area of the base is πr^2.

We'll also often consider the surface area of a three-dimensional container. For a rectangular box with side lengths of l, w, and h, its surface area consists of 3 pairs of rectangles: the top and bottom, each of area lw, the two sides that are the front and back when we look right at the box, each of area lh, and the remaining two sides of area wh. Thus the total surface area of the box is

$$SA = 2lw + 2lh + 2wh. \tag{5.3.3}$$

For a circular cylinder, its surface area is the sum of the areas of the top and bottom (πr^2 each), plus the area of the "sides". If we think of cutting the cylinder vertically and unfurling it, the resulting figure is a rectangle whose dimensions are the height of the cylinder, h, by the circumference of the base, $2\pi r$. The rectangle's area is therefore $2\pi r \cdot h$, and hence the total surface area of a cylinder is

$$SA = 2\pi r^2 + 2\pi rh. \qquad (5.3.4)$$

Each of the volume and surface area equations (Equation (5.3.1), Equation (5.3.2), Equation (5.3.3), and Equation (5.3.4)) involve only multiplication and addition, and thus have the potential to result in polynomial functions. At present, however, each of these equations involves at least two variables. The inclusion of additional constraints can enable us to use these formulas to generate polynomial functions of a single variable.

> **Activity 5.3.2.** According to a shipping company's regulations, the girth plus the length of a parcel they transport for their lowest rate may not exceed 120 inches, where by *girth* we mean the perimeter of the smallest end.
>
>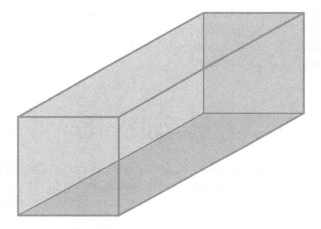
>
> **Figure 5.3.3:** A rectangular parcel with a square end.
>
> Suppose that we want to ship a parcel that has a square end of width x and an overall length of y, both measured in inches.
>
> a. Label the provided picture, using x for the length of each side of the square end, and y for the other edge of the package.
>
> b. How does the length plus girth of 120 inches result in an equation (often called a *constraint* equation) that relates x and y? Explain, and state the equation.
>
> c. Solve the equation you found in (b) for one of the variables present.
>
> d. Hence determine the volume, V, of the package as a function of a single variable.
>
> e. What is the domain of the function V in the context of the physical setting of this problem? (Hint: what's the maximum value of x? the maximum value of y?)

Activity 5.3.3. Suppose that we want to construct a cylindrical can using 60 square inches of material for the surface of the can. In this context, how does the can's volume depend on the radius we choose?

Let the cylindrical can have base radius r and height h.

a. Use the formula for the surface area of a cylinder and the given constraint that the can's surface area is 60 square inches to write an equation that connects the radius r and height h.

b. Solve the equation you found in (a) for h in terms of r.

c. Recall that the volume of a cylinder is $V = \pi r^2 h$. Use your work in (b) to write V as a function of r only and simplify the formula as much as possible.

d. Hence determine the volume, V, of the package as a function of a single variable.

e. What is the domain of the function V in the context of the physical setting of this problem? (Hint: how does the constraint on surface area provide a largest possible value for r? Think about the maximum area that can be allocated to the top and bottom of the can.)

5.3.2 Other applications of polynomial functions

A different use of polynomial functions arises with *Bezier curves*. The most common type of Bezier curve used in applications is the cubic Bezier curve, which is a curve given parametrically by a formula of the form $(x(t), y(t))$, where

$$x(t) = (1 - t)^3 x_0 + 3(1 - t^2)t x_1 + 3(1 - t^2)t x_2 + t^3 x_3$$

and

$$y(t) = (1 - t)^3 y_0 + 3(1 - t^2)t y_1 + 3(1 - t)t^2 y_2 + t^3 y_3).$$

The curve passes through the points $A = (x_0, y_0)$ and $B = (x_3, y_3)$ and the points $C = (x_1, y_1)$ and $D = (x_2, y_2)$ are called *control points*. At http://gvsu.edu/s/0zC, you can explore the effects of moving the control points (in gray) and the points on the curve (in black) to generate different curves in the plane, similar to the one shown in Figure 5.3.4.

The main issue to realize is that the form of the curve depends on a special family of cubic polynomials:

$$(1 - t)^3, 3(1 - t^2)t, 3(1 - t^2)t, \text{ and } t^3.$$

These four cubic polynomials play a key role in graphic design and are used in all sorts of important ways, including in font design, as seen in Figure 5.3.5.

Another important application of polynomial functions is found in how they can be used to approximate the sine and cosine functions.

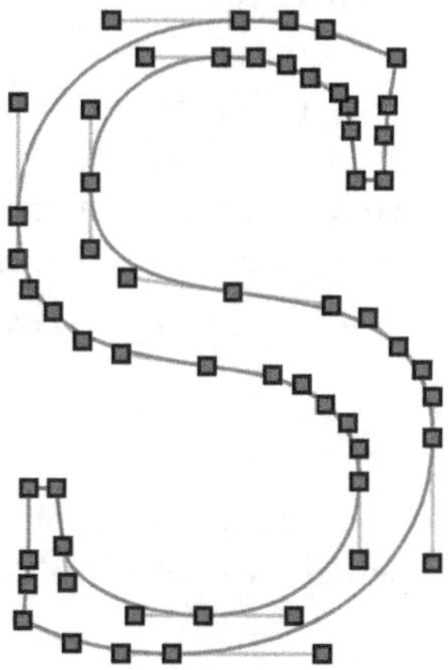

Figure 5.3.5: The letter S in Palatino font, generated by Bezier curves.

Figure 5.3.4: A cubic Bezier curve with control points in gray.

Activity 5.3.4. We understand the theoretical rule behind the function $f(t) = \sin(t)$: given an angle t in radians, $\sin(t)$ measures the value of the y-coordinate of the corresponding point on the unit circle. For special values of t, we have determined the exact value of $\sin(t)$. For example, $\sin(\frac{\pi}{3}) = \frac{\sqrt{3}}{2}$. But note that we don't have a *formula* for $\sin(t)$. Instead, we use a button on our calculator or command on our computer to find values like "sin(1.35)." It turns out that a combination of calculus and polynomial functions explains how computers determine values of the sine function.

At http://gvsu.edu/s/0zA, you'll find a *Desmos* worksheet that has the sine function already defined, along with a sequence of polynomials labeled $T_1(x)$, $T_3(x)$, $T_5(x)$, $T_7(x)$, …. You can see these functions' graphs by clicking on their respective icons.

a. For what values of x does it appear that $\sin(x) \approx T_1(x)$?

b. For what values of x does it appear that $\sin(x) \approx T_3(x)$?

c. For what values of x does it appear that $\sin(x) \approx T_5(x)$?

d. What overall trend do you observe? How good is the approximation generated by $T_{19}(x)$?

e. In a new *Desmos* worksheet, plot the function $y = \cos(x)$ along with the following functions: $P_2(x) = 1 - \frac{x^2}{2!}$ and $P_4(x) = 1 - \frac{x^2}{2!} + \frac{x^4}{4!}$. Based on the patterns with the coefficients in the polynomials approximating $\sin(x)$ and the polynomials P_2 and P_4 here, conjecture formulas for P_6, P_8, and P_{18} and plot them. How well can we approximate $y = \cos(x)$ using polynomials?

5.3.3 Summary

- Polynomials arise naturally in the study of problems involving the volume and surface area of three-dimensional containers such as boxes and cylinders because these formulas fundamentally involve sums and products of variables. For instance, the volume of a cylinder is $V = \pi r^2 h$. In the presence of a surface area constraint that tells us that $h = \frac{100 - 2\pi r^2}{2\pi r}$, it follows that

$$V = \pi r^2 \frac{100 - 2\pi r^2}{2\pi r} = r(50 - \pi r^2),$$

  which is a cubic polynomial.

- Polynomial functions be used to approximate non-polynomial curves and functions in many different ways. One example is found in cubic Bezier curves which use a collection of *control points* to enable the user to manipulate curves to pass through select points in such a way that the curve first travels in a certain direction. Another example is in the remarkable approximation of non-polynomial functions like the sine function, as given by

$$\sin(x) \approx x - \frac{1}{3!}x^3 + \frac{1}{5!}x^5 - \frac{1}{7!}x^7,$$

  where the approximation is good for x-values near $x = 0$.

5.3.4 Exercises

1. You wish to pack a cardboard box inside a wooden crate. In order to have room for the packing materials, you need to leave a 0.5 ft space around the front, back, and sides of the box, and a 1 ft space around the top and bottom of the box.

If the cardboard box is x feet long, $(x+2)$ feet wide, and $(x-1)$ feet deep, find a formula in terms of x for the amount of packing material, M, needed.

2. An open-top box is to be constructed from a 6 in by 14 in rectangular sheet of tin by cutting out squares of equal size at each corner, then folding up the resulting flaps. Let x denote the length of the side of each cut-out square. Assume negligible thickness.

(a) Find a formula for the volume, V, of the box as a function of x.

(b) For what values of x does the formula from part (a) make sense in the context of the problem?

(c) On a separate piece of paper, sketch a graph of the volume function.

(d) What, approximately, is the maximum volume of the box?

3. An open triangular trough, as pictured in Figure 5.3.6 is being contructed from aluminum. The trough is to have equilateral triangular ends of side length s and a length of l. We want the trough to used a fixed 100 square feet of aluminum.

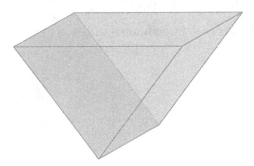

Figure 5.3.6: A triangular trough.

a. What is the area of one of the equilateral triangle ends as a function of s?

b. Recall that for an object with constant cross-sectional area, its volume is the area of one of those cross-sections times its height (or length). Hence determine a formula for the volume of the trough that depends on s and l.

c. Find a formula involving s and l for the surface area of the trough.

d. Use the constraint that we have 100 square feet of available aluminum to generate an equation that connects s and l and hence solve for l in terms of s.

e. Use your work in (d) and (b) to express the volume of the trough, V, as a function of l only.

f. What is the domain of the function V in the context of the situation being modeled? Why?

4. A rectangular box is being constructed so that its base is twice as long as it is wide. In addition, the base and top of the box cost $2 per square foot while the sides cost $1.50 per square foot. If we only want to spend $10 on materials for the box, how can we write the box's volume as a function of a single variable? What is the domain of this volume function? (Hint: first find the box's surface area in terms of two variables, and then find an expression for the cost of the box in terms of those same variables. Use the fact that cost is constrained to solve for one variable in terms of another.)

5. Suppose that we want a cylindrical barrel to hold 8 cubic feet of volume. Let the barrel have radius r and height h, each measured in feet. How can we write the surface area, A, of the barrel solely as a function of r?

a. Draw several possible pictures of how the barrel might look. For instance, what if the radius is very small? How will the height appear in comparison? Likewise, what happens if the height is very small?

b. Use the fact that volume is fixed at 8 cubic feet to state a constraint equation and solve that equation for h in terms of r.

c. Recall that the surface area of a cylinder is $A = 2\pi r^2 + 2\pi rh$. Use your work in (c) to write A as a function of only r.

    d. What is the domain of A? Why?

    e. Explain why A is *not* a polynomial function of r.

5.4 Rational Functions

Motivating Questions

- What is a rational function?

- How can we determine key information about a rational function from its algebraic structure?

- Why are rational functions important?

The average rate of change of a function on an interval always involves a ratio. Indeed, for a given function f that interests us near $t = 2$, we can investigate its average rate of change on intervals near this value by considering

$$AV_{[2,2+h]} = \frac{f(2+h) - f(2)}{h}.$$

Suppose, for instance, that f meausures the height of a falling ball at time t and is given by $f(t) = -16t^2 + 32t + 48$, which happens to be a polynomial function of degree 2. For this particular function, its average rate of change on $[1, 1 + h]$ is

$$\begin{aligned} AV_{[2,2+h]} &= \frac{f(2+h) - f(2)}{h} \\ &= \frac{-16(2+h)^2 + 32(2+h) + 48 - (-16 \cdot 4 + 32 \cdot 2 + 48)}{h} \\ &= \frac{-64 - 64h - 16h^2 + 64 + 32h + 48 - (48)}{h} \\ &= \frac{-64h - 16h^2}{h}. \end{aligned}$$

Structurally, we observe that $AV_{[2,2+h]}$ is a ratio of the two functions $-64h - 16h^2$ and h. Moreover, both the numerator and the denominator of the expression are themselves polynomial functions of the variable h. Note that we may be especially interested in what occurs as $h \to 0$, as these values will tell us the average velocity of the moving ball on shorter and shorter time intervals starting at $t = 2$. At the same time, $AV_{[2,2+h]}$ is not defined for $h = 0$.

Ratios of polynomial functions arise in several different important circumstances. Sometimes we are interested in what happens when the denominator approaches 0, which makes the overall ratio undefined. In other situations, we may want to know what happens in the long term and thus consider what happens when the input variable increases without bound.

> **Preview Activity 5.4.1.** A drug company[1] estimates that to produce a new drug, it will cost $5 million in startup resources, and that once they reach production, each gram of the drug will cost $2500 to make.
>
> a. Determine a formula for a function $C(q)$ that models the cost of producing q grams of the drug. What familiar kind of function is C?

b. The drug company needs to sell the drug at a price of more than $2500 per gram in order to at least break even. To investigate how they might set prices, they first consider what their *average* cost per gram is. What is the *total* cost of producing 1000 grams? What is the *average* cost per gram to produce 1000 grams?

c. What is the total cost of producing 10000 grams? What is the average cost per gram to produce 10000 grams?

d. Our computations in (b) and (c) naturally lead us to define the "average cost per gram" function, $A(q)$, whose output is the average cost of producing q grams of the drug. What is a formula for $A(q)$?

e. Explain why another formula for A is $A(q) = 2500 + \frac{5000000}{q}$.

f. What can you say about the long-range behavior of A? What does this behavior mean in the context of the problem?

5.4.1 Long-range behavior of rational functions

The functions $AV_{[2,2+h]} = \frac{-64h-16h^2}{h}$ and $A(q) = \frac{5000000+2500q}{q}$ are both examples of rational functions, since each is a ratio of polynomial functions. Formally, we have the following definition.

Definition 5.4.1 A function r is **rational** provided that it is possible to write r as the ratio of two polynomials, p and q. That is, r is rational provided that for some polynomial functions p and q, we have

$$r(x) = \frac{p(x)}{q(x)}.$$

◊

Like with polynomial functions, we are interested in such natural questions as

- What is the long range behavior of a given rational function?

- What is the domain of a given rational function?

- How can we determine where a given rational function's value is 0?

We begin by focusing on the long-range behavior of rational functions. It's important first to recall our earlier work with power functions of the form $p(x) = x^{-n}$ where $n = 1, 2, \ldots$. For such functions, we know that $p(x) = \frac{1}{x^n}$ where $n > 0$ and that

$$\lim_{x \to \infty} \frac{1}{x^n} = 0$$

since x^n increases without bound as $x \to \infty$. The same is true when $x \to -\infty$: $\lim_{x \to -\infty} \frac{1}{x^n} = 0$. Thus, any time we encounter a quantity such as $\frac{1}{x^3}$, this quantity will approach 0 as x

---

[1]This activity is based on p. 457ff in *Functions Modeling Change*, by Connally et al.

increases without bound, and this will also occur for any constant numerator. For instance,

$$\lim_{x \to \infty} \frac{2500}{x^2} = 0$$

since 2500 times a quantity approaching 0 will still approach 0 as x increases.

Activity 5.4.2. Consider the rational function $r(x) = \frac{3x^2 - 5x + 1}{7x^2 + 2x - 11}$.

Observe that the largest power of x that's present in $r(x)$ is x^2. In addition, because of the dominant terms of $3x^2$ in the numerator and $7x^2$ in the denominator, both the numerator and denominator of r increase without bound as x increases without bound. In order to understand the long-range behavior of r, we choose to write the function in a different algebraic form.

a. Note that we can multiply the formula for r by the form of 1 given by $1 = \frac{\frac{1}{x^2}}{\frac{1}{x^2}}$.

Do so, and distribute and simplify as much as possible in both the numerator and denominator to write r in a different algebraic form.

b. Having rewritten r, we are in a better position to evaluate $\lim_{x \to \infty} r(x)$. Using our work from (a), we have

$$\lim_{x \to \infty} r(x) = \lim_{x \to \infty} \frac{3 - \frac{5}{x} + \frac{1}{x^2}}{7 + \frac{2}{x} - \frac{11}{x^2}}.$$

What is the exact value of this limit and why?

c. Next, determine

$$\lim_{x \to -\infty} r(x) = \lim_{x \to -\infty} \frac{3 - \frac{5}{x} + \frac{1}{x^2}}{7 + \frac{2}{x} - \frac{11}{x^2}}.$$

d. Use *Desmos* to plot r on the interval $[-10, 10]$. In addition, plot the horizontal line $y = \frac{3}{7}$. What is the meaning of the limits you found in (b) and (c)?

Activity 5.4.3. Let $s(x) = \frac{3x - 5}{7x^2 + 2x - 11}$ and $u(x) = \frac{3x^2 - 5x + 1}{7x + 2}$. Note that both the numerator and denominator of each of these rational functions increases without bound as $x \to \infty$, and in addition that x^2 is the highest order term present in each of s and u.

a. Using a similar algebraic approach to our work in Activity 5.4.2, multiply $s(x)$ by $1 = \frac{\frac{1}{x^2}}{\frac{1}{x^2}}$ and hence evaluate

$$\lim_{x \to \infty} \frac{3x - 5}{7x^2 + 2x - 11}.$$

What value do you find?

b. Plot the function $y = s(x)$ on the interval $[-10, 10]$. What is the graphical meaning of the limit you found in (a)?

c. Next, use appropriate algebraic work to consider $u(x)$ and evaluate

$$\lim_{x \to \infty} \frac{3x^2 - 5x + 1}{7x + 2}.$$

What do you find?

d. Plot the function $y = u(x)$ on the interval $[-10, 10]$. What is the graphical meaning of the limit you computed in (c)?

We summarize and generalize the results of Activity 5.4.2 and Activity 5.4.3 as follows.

The long-term behavior of a rational function.

Let p and q be polynomial functions so that $r(x) = \frac{p(x)}{q(x)}$ is a rational function. Suppose that p has degree n with leading term $a_n x^n$ and q has degree m with leading term $b_m x^m$ for some nonzero constants a_n and b_m. There are three possibilities ($n < m$, $n = m$, and $n > m$) that result in three different behaviors of r:

a. if $n < m$, then the degree of the numerator is less than the degree of the denominator, and thus

$$\lim_{n \to \infty} r(x) = \lim_{n \to \infty} \frac{a_n x^n + \cdots + a_0}{b_m x^m + \cdots + b_0} = 0,$$

which tells us that $y = 0$ is a horizontal asymptote of r;

b. if $n = m$, then the degree of the numerator equals the degree of the denominator, and thus

$$\lim_{n \to \infty} r(x) = \lim_{n \to \infty} \frac{a_n x^n + \cdots + a_0}{b_n x^n + \cdots + b_0} = \frac{a_n}{b_n},$$

which tells us that $y = \frac{a_n}{b_n}$ (the ratio of the coefficients of the highest order terms in p and q) is a horizontal asymptote of r;

c. if $n > m$, then the degree of the numerator is greater than the degree of the denominator, and thus

$$\lim_{n \to \infty} r(x) = \lim_{n \to \infty} \frac{a_n x^n + \cdots + a_0}{b_m x^m + \cdots + b_0} = \pm\infty,$$

(where the sign of the limit depends on the signs of a_n and b_m) which tells us that r is does not have a horizontal asymptote.

In both situations (a) and (b), the value of $\lim_{x \to -\infty} r(x)$ is identical to $\lim_{x \to \infty} r(x)$.

5.4.2 The domain of a rational function

Because a rational function can be written in the form $r(x) = \frac{p(x)}{q(x)}$ for some polynomial functions p and q, we have to be concerned about the possibility that a rational function's denominator is zero. Since polynomial functions always have their domain as the set of all real numbers, it follows that any rational function is only undefined at points where its denominator is zero.

> **The domain of a rational function.**
>
> Let p and q be polynomial functions so that $r(x) = \frac{p(x)}{q(x)}$ is a rational function. The domain of r is the set of all real numbers except those for which $q(x) = 0$.

Example 5.4.2 Determine the domain of the function $r(x) = \frac{5x^3+17x^2-9x+4}{2x^3-6x^2-8x}$.

Solution. To find the domain of any rational function, we need to determine where the denominator is zero. The best way to find these values exactly is to factor the denominator. Thus, we observe that

$$2x^3 - 6x^2 - 8x = 2x(x^2 - 3x - 4) = 2x(x + 1)(x - 4).$$

By the Zero Product Property, it follows that the denominator of r is zero at $x = 0$, $x = -1$, and $x = 4$. Hence, the domain of r is the set of all real numbers except $-1, 0$, and 4. □

We note that when it comes to determining the domain of a rational function, the numerator is irrelevant: all that matters is where the denominator is 0.

> **Activity 5.4.4.** Determine the domain of each of the following functions. In each case, write a sentence to accurately describe the domain.
>
> a. $f(x) = \frac{x^2 - 1}{x^2 + 1}$
>
> b. $g(x) = \frac{x^2 - 1}{x^2 + 3x - 4}$
>
> c. $h(x) = \frac{1}{x} + \frac{1}{x - 1} + \frac{1}{x - 2}$
>
> d. $j(x) = \frac{(x + 5)(x - 3)(x + 1)(x - 4)}{(x + 1)(x + 3)(x - 5)}$
>
> e. $k(x) = \frac{2x^2 + 7}{3x^3 - 12x}$
>
> f. $m(x) = \frac{5x^2 - 45}{7(x - 2)(x - 3)^2(x^2 + 9)(x + 1)}$

5.4.3 Applications of rational functions

Rational functions arise naturally in the study of the average rate of change of a polynomial function, leading to expressions such as

$$AV_{[2,2+h]} = \frac{-64h - 16h^2}{h}.$$

We will study several subtle issues that correspond to such functions further in Section 5.5. For now, we will focus on a different setting in which rational functions play a key role.

In Section 5.3, we encountered a class of problems where a key quantity was modeled by a polynomial function. We found that if we considered a container such as a cylinder with fixed surface area, then the volume of the container could be written as a polynomial of a single variable. For instance, if we consider a circular cylinder with surface area 10 square feet, then we know that

$$S = 10 = 2\pi r^2 + 2\pi rh$$

and therefore $h = \frac{10-2\pi r^2}{2\pi r}$. Since the cylinder's volume is $V = \pi r^2 h$, it follows that

$$V = \pi r^2 h = \pi r^2 \left(\frac{10 - 2\pi r^2}{2\pi r} \right) = r(10 - 2\pi r^2),$$

which is a polynomial function of r.

What happens if we instead fix the volume of the container and ask about how surface area can be written as a function of a single variable?

Example 5.4.3 Suppose we want to construct a circular cylinder that holds 20 cubic feet of volume. How much material does it take to build the container? How can we state the amount of material as a function of a single variable?

Solution. Neglecting any scrap, the amount of material it takes to construct the container is its surface area, which we know to be

$$S = 2\pi r^2 + 2\pi rh.$$

Because we want the volume to be fixed, this results in a constraint equation that enables us to relate r and h. In particular, since

$$V = 20 = \pi r^2 h,$$

it follows that we can solve for h and get $h = \frac{20}{\pi r^2}$. Substituting this expression for h in the equation for surface area, we find that

$$S = 2\pi r^2 + 2\pi r \cdot \frac{20}{\pi r^2} = 2\pi r^2 + \frac{40}{r}.$$

Getting a common denominator, we can also write S in the form

$$S(r) = \frac{2\pi r^3 + 40}{r}$$

and thus we see that S is a rational function of r. Because of the physical context of the problem and the fact that the denominator of S is r, the domain of S is the set of all positive real numbers. □

Activity 5.4.5. Suppose that we want to build an open rectangular box (that is, without a top) that holds 15 cubic feet of volume. If we want one side of the base to be twice as long as the other, how does the amount of material required depend on the shorter side of the base? We investigate this question through the following sequence of prompts.

    a. Draw a labeled picture of the box. Let x represent the shorter side of the base and h the height of the box. What is the length of the longer side of the base in terms of x?

    b. Use the given volume constraint to write an equation that relates x and h, and solve the equation for h in terms of x.

    c. Determine a formula for the surface area, S, of the box in terms of x and h.

    d. Using the constraint equation from (b) together with your work in (c), write surface area, S, as a function of the single variable x.

    e. What type of function is S? What is its domain?

    f. Plot the function S using *Desmos*. What appears to be the least amount of material that can be used to construct the desired box that holds 15 cubic feet of volume?

5.4.4 Summary

- A rational function is a function whose formula can be written as the ratio of two polynomial functions. For instance, $r(x) = \frac{7x^3-5x+16}{-4x^4+2x^3-11x+3}$ is a rational function.

- Two aspects of rational functions are straightforward to determine for any rational function. Given $r(x) = \frac{p(x)}{q(x)}$ where p and q are polynomials, the domain of r is the set of all real numbers except any values of x for which $q(x) = 0$. In addition, we can determine the long-range behavior of r by examining the highest order terms in p and q:

    ○ if the degree of p is less than the degree of q, then r has a horizontal asymptote at $y = 0$;

    ○ if the degree of p equals the degree of q, then r has a horizontal asymptote at $y = \frac{a_n}{b_n}$, where a_n and b_n are the leading coefficients of p and q respectively;

    ○ and if the degree of p is greater than the degree of q, then r does not have a horizontal asymptote.

- Two reasons that rational functions are important are that they arise naturally when we consider the average rate of change on an interval whose length varies and when we consider problems that relate the volume and surface area of three-dimensional containers when one of those two quantities is constrained.

5.4.5 Exercises

1. Find the horizontal asymptote, if it exists, of the rational function below.

$$g(x) = \frac{(-1-x)(-7-2x)}{2x^2 + 1}$$

2. Compare and discuss the long-run behaviors of the functions below. In each blank, enter either the constant or the polynomial that the rational function behaves like as $x \to \pm\infty$:

$$f(x) = \frac{x^3 + 3}{x^3 - 8}, \ g(x) = \frac{x^2 + 3}{x^3 - 8}, \text{ and } h(x) = \frac{x^4 + 3}{x^3 - 8}$$

$f(x)$ will behave like the function $y =$ _____ as $x \to \pm\infty$.

$g(x)$ will behave like the function $y =$ _____ as $x \to \pm\infty$.

$h(x)$ will behave like the function $y =$ _____ as $x \to \pm\infty$.

3. Let $r(x) = \dfrac{p(x)}{q(x)}$, where p and q are polynomials of degrees m and n respectively.

(a) If $r(x) \to 0$ as $x \to \infty$, then

☐ $m > n$ ☐ $m = n$ ☐ $m < n$ ☐ None of the above

(b) If $r(x) \to k$ as $x \to \infty$, with $k \neq 0$, then

☐ $m < n$ ☐ $m > n$ ☐ $m = n$ ☐ None of the above

4. Find all zeros and vertical asymptotes of the rational function

$$f(x) = \frac{x + 6}{(x + 9)^2}.$$

(a) The function has zero(s) at $x =$ _____

(b) The function has vertical asymptote(s) at $x =$ _____

(c) The function's long-run behavior is that $y \to$ _____ as $x \to \pm\infty$

(d) On a piece of paper, sketch a graph of this function without using your calculator.

5. Find all zeros and vertical asymptotes of the rational function

$$f(x) = \frac{x^2 - 16}{-x^3 - 16x^2}.$$

(a) The function has x-intercept(s) at $x =$ _____

(b) The function has y-intercept(s) at $y =$ _____

(c) The function has vertical asymptote(s) when $x =$ _____

(d) The function has horizontal asymptote(s) when $y =$ _____

6. Using the graph of the rational function $y = f(x)$ given in the figure below, evaluate the limits.

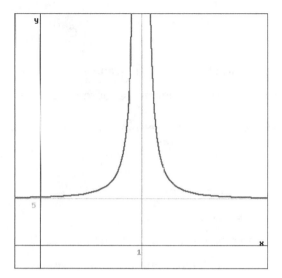

(a) $\lim\limits_{x \to \infty} f(x)$

(b) $\lim\limits_{x \to -\infty} f(x)$

(c) $\lim\limits_{x \to 1^+} f(x)$

(d) $\lim\limits_{x \to 1^-} f(x)$

7. The graph below is a vertical and/or horizontal shift of $y = 1/x$ (assume no reflections or compression/expansions have been applied).

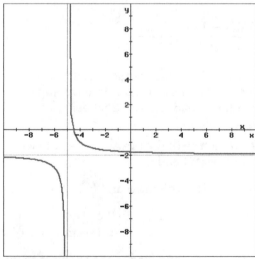

(a) The graph's equation can be written in the form

$$f(x) = \frac{1}{x + A} + B$$

for constants A and B. Based on the graph above, find the values for A and B.

(b) Now take your formula from part (a) and write it as the ratio of two linear polynomials of the form,

$$f(x) = \frac{Mx + C}{x + D}$$

for constants M, C, and D. What are the values of M, C, and D?

(c) Find the exact values of the coordinates of the x- and y-intercepts of the graph.

8. Find all zeros and vertical asymptotes of the rational function

$$f(x) = \frac{x^2 - 1}{x^2 + 1}.$$

(a) The function has x-intercept(s) at $x =$ _____

(b) The function has y-intercept(s) at $y =$ _____

(c) The function has vertical asymptote(s) when $x =$ _____

(d) The function has horizontal asymptote(s) when $y =$ _____

9. For each rational function below, determine the function's domain as well as the exact value of any horizontal asymptote.

a. $f(x) = \dfrac{17x^2 + 34}{19x^2 - 76}$

b. $g(x) = \dfrac{29}{53} + \dfrac{1}{x - 2}$

c. $h(x) = \dfrac{4 - 31x}{11x - 7}$

d. $r(x) = \dfrac{151(x - 4)(x + 5)^2(x - 2)}{537(x + 5)(x + 1)(x^2 + 1)(x - 15)}$

10. A rectangular box is being constructed so that its base is 1.5 times as long as it is wide. In addition, suppose that material for the base and top of the box costs $3.75 per square foot, while material for the sides costs $2.50 per square foot. Finally, we want the box to hold 8 cubic feet of volume.

a. Draw a labeled picture of the box with x as the length of the shorter side of the box's base and h as its height.

b. Determine a formula involving x and h for the total surface area, S, of the box.

c. Use your work from (b) along with the given information about cost to determine a formula for the total cost, C, oif the box in terms of x and h.

d. Use the volume constraint given in the problem to write an equation that relates x and h, and solve that equation for h in terms of x.

e. Combine your work in (c) and (d) to write the cost, C, of the box as a function solely of x.

f. What is the domain of the cost function? How does a graph of the cost function appear? What does this suggest about the ideal box for the given constraints?

11. A cylindrical can is being constructed so that its volume is 16 cubic inches. Suppose that material for the lids (the top and bottom) cost \$0.11 per square inch and material for the "side" of the can costs \$0.07 per square inch. Determine a formula for the total cost of the can as a function of the can's radius. What is the domain of the function and why?

Hint. You may find it helpful to ask yourself a sequence of questions like those stated in Exercise 10).

5.5 Key features of rational functions

Motivating Questions

- What does it mean to say that a rational function has a "hole" at a certain point, and what algebraic structure leads to such behavior?

- How do we determine where a rational function has zeros and where it has vertical asymptotes?

- What does a sign chart reveal about the behavior of a rational function and how do we develop a sign chart from a given formula?

Because any rational function is the ratio of two polynomial functions, it's natural to ask questions about rational functions similar to those we ask about polynomials. With polynomials, it is often helpful to know where the function's value is zero. In a rational function $r(x) = \frac{p(x)}{q(x)}$, we are curious to know where both $p(x) = 0$ and where $q(x) = 0$.

Connected to these questions, we want to understand both where a rational function's output value is zero, as well as where the function is undefined. In addition, from the behavior of simple rational power functions such as $\frac{1}{x}$, we expect that rational functions may not only have horizontal asymptotes (as investigated in Section 5.4), but also vertical asymptotes. At first glance, these questions about zeros and vertical asymptotes of rational functions may appear to be elementary ones whose answers simply depend on where the numerator and denominator of the rational function are zero. But in fact, rational functions often admit very subtle behavior that can escape the human eye and the graph generated by a computer.

Preview Activity 5.5.1. Consider the rational function $r(x) = \frac{x^2-1}{x^2-3x-4}$, and let $p(x) = x^2 - 1$ (the numerator of $r(x)$) and $q(x) = x^2 - 3x - 4$ (the denominator of $r(x)$).

    a. Reasoning algebraically, for what values of x is $p(x) = 0$?

    b. Again reasoning algebraically, for what values of x is $q(x) = 0$?

    c. Define $r(x)$ in *Desmos*, and evaluate the function appropriately to find numerical values for the output of r and hence complete the following tables.

x	$r(x)$	x	$r(x)$	x	$r(x)$
4.1		1.1		−1.1	
4.01		1.01		−1.01	
4.001		1.001		−1.001	
3.9		0.9		−0.9	
3.99		0.99		−0.99	
3.999		0.999		−0.999	

    d. Why does r behave the way it does near $x = 4$? Explain by describing the behavior of the numerator and denominator.

   e. Why does r behave the way it does near $x = 1$? Explain by describing the behavior of the numerator and denominator.

   f. Why does r behave the way it does near $x = -1$? Explain by describing the behavior of the numerator and denominator.

   g. Plot r in *Desmos*. Is there anything surprising or misleading about the graph that *Desmos* generates?

5.5.1 When a rational function has a "hole"

Two important features of any rational function $r(x) = \frac{p(x)}{q(x)}$ are any zeros and vertical asymptotes the function may have. These aspects of a rational function are closely connected to where the numerator and denominator, respectively, are zero. At the same time, a subtle related issue can lead to radically different behavior. To understand why, we first remind ourselves of a few key facts about fractions that involve 0. Because we are working with a function, we'll think about fractions whose numerator and denominator are approaching particular values.

If the numerator of a fraction approaches 0 while the denominator approaches a nonzero value, then the overall fraction values will approach zero. For instance, consider the sequence of values

$$\frac{0.1}{0.9} = 0.111111\cdots , \quad \frac{0.01}{0.99} = 0.010101\cdots , \quad \frac{0.001}{0.999} = 0.001001\cdots .$$

Because the numerator gets closer and closer to 0 and the denominator stays away from 0, the quotients tend to 0.

Similarly, if the denominator of a fraction approaches 0 while the numerator approaches a nonzero value, then the overall fraction increases without bound. If we consider the reciprocal values of the sequence above, we see that

$$\frac{0.9}{0.1} = 9, \quad \frac{0.99}{0.01} = 99, \quad \frac{0.999}{0.001} = 999.$$

Since the denominator gets closer and closer to 0 and the numerator stays away from 0, the quotients increase without bound.

These two behaviors show how the zeros and vertical asympototes of a rational function $r(x) = \frac{p(x)}{q(x)}$ arise: where the numerator $p(x)$ is zero and the denominator $q(x)$ is nonzero, the function r will have a zero; and where $q(x)$ is zero and $p(x)$ is nonzero, the function will have a vertical asymptote. What we must be careful of is the special situation where *both* the numerator $p(x)$ and $q(x)$ are simultaneously zero. Indeed, if the numerator and denominator of a fraction both approach 0, different behavior can arise. For instance, consider the sequence

$$\frac{0.2}{0.1} = 2, \quad \frac{0.02}{0.01} = 2, \quad \frac{0.002}{0.001} = 2.$$

In this situation, both the numerator and denominator are approaching 0, but the overall fraction's value is always 2. This is very different from the two sequences we considered above. In Example 5.5.1, we explore similar behavior in the context of a particular rational function.

Example 5.5.1 Consider the rational function $r(x) = \frac{x^2-1}{x^2-3x-4}$ from Preview Activity 5.5.1, whose numerator is $p(x) = x^2 - 1$ and whose denominator is $q(x) = x^2 - 3x - 4$. Explain why the graph of r generated by *Desmos* or another computational device is incorrect, and also identify the locations of any zeros and vertical asymptotes of r.

Solution.   It is helpful with any rational function to factor the numerator and denominator. We note that $p(x) = x^2 - 1 = (x-1)(x+1)$ and $q(x) = x^2 - 3x - 4 = (x+1)(x-4)$. The domain of r is thus the set of all real numbers except $x = -1$ and $x = 4$, the set of all points where $q(x) \neq 0$.

Knowing that r is not defined at $x = -1$, it is natural to study the graph of r near that value. Plotting the function in *Desmos*, we get a result similar to the one shown in Figure 5.5.2, which appears to show no unusual behavior at $x = -1$, and even that $r(-1)$ is defined. If we zoom in on that point, as shown in Figure 5.5.3, the technology still fails to visually demonstrate the fact that $r(-1)$ is not defined. This is because graphing utilities sample functions at a finite number of points and then connect the resulting dots to generate the curve we see.

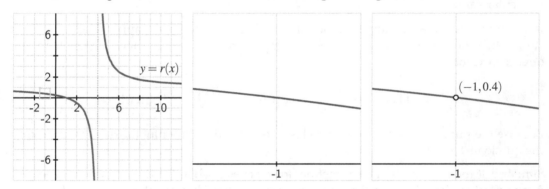

Figure 5.5.2: A plot of $r(x) = \frac{x^2-1}{x^2-3x-4}$.

Figure 5.5.3: Zooming in on $r(x)$ near $x = -1$.

Figure 5.5.4: How the graph of $r(x)$ should actually appear near $x = -1$.

We know from our algebraic work with the denominator, $q(x) = (x + 1)(x - 4)$, that r is not defined at $x = -1$. While the denominator q gets closer and closer to 0 as x approaches -1, so does the numerator, since $p(x) = (x - 1)(x + 1)$. If we consider values close but not equal to $x = -1$, we see results in Table 5.5.5.

x	-1.1	-1.01	-1.001
$r(x)$	$\frac{0.21}{0.51} \approx 0.4118$	$\frac{0.0201}{0.0501} \approx 0.4012$	$\frac{0.002001}{0.005001} \approx 0.4001$

x	-0.9	-0.99	-0.999
$r(x)$	$\frac{-0.19}{-0.49} \approx 0.3878$	$\frac{-0.0199}{-0.0499} \approx 0.3989$	$\frac{-0.001999}{-0.004999} \approx 0.3999$

Table 5.5.5: Values of $r(x) = \frac{x^2-1}{x^2-3x-4}$ near $x = -1$.

In the table, we see that both the numerator and denominator get closer and closer to 0 as x gets closer and closer to -1, but that their quotient appears to be getting closer and closer to $y = 0.4$. Indeed, we see this behavior in the graph of r, though the graphing utility misses the fact that $r(-1)$ is actually not defined. A precise graph of r near $x = -1$ should look like the one presented in Figure 5.5.4, where we see an open circle at the point $(-1, 0.4)$ that demonstrates that $r(-1)$ is not defined, and that r does not have a vertical asymptote or zero at $x = -1$.

Finally, we also note that $p(1) = 0$ and $q(1) = -6$, so at $x = 1$, $r(x)$ has a zero (its numerator is zero and its denominator is not). In addition, $q(4) = 0$ and $p(4) = 15$ (its denominator is zero and its numerator is not), so $r(x)$ has a vertical asymptote at $x = 4$. These features are accurately represented by the original *Desmos* graph shown in Figure 5.5.2. □

In the situation where a rational function is undefined at a point but does not have a vertical asymptote there, we'll say that the graph of the function has a **hole**. In calculus, we use limit notation to identify a hole in a function's graph. Indeed, having shown in Example 5.5.1 that the value of $r(x)$ gets closer and closer to 0.4 as x gets closer and closer to -1, we naturally write $\lim_{x \to -1} r(x) = 0.4$ as a shorthand way to represent the behavior of r (similar to how we've written limits involving ∞). This fact, combined with $r(-1)$ being undefined, tells us that near $x = -1$ the graph approaches a value of 0.4 but has to have a hole at the point $(-1, 0.4)$, as shown in Figure 5.5.4. Because we'll encounter similar behavior with other functions, we formally define limit notation as follows.

Definition 5.5.6 Let a and L be finite real numbers, and let r be a function defined near $x = a$, but not necessarily at $x = a$ itself. If we can make the value of $r(x)$ as close to the number L as we like by taking x sufficiently close (but not equal) to a, then we write

$$\lim_{x \to a} r(x) = L$$

and say that "the limit of r as x approaches a is L". ◊

The key observations regarding zeros, vertical asymptotes, and holes in Example 5.5.1 apply to any rational function.

Features of a rational function.

Let $r(x) = \frac{p(x)}{q(x)}$ be a rational function.

- If $p(a) = 0$ and $q(a) \neq 0$, then $r(a) = 0$, so r has a zero at $x = a$.

- If $q(a) = 0$ and $p(a) \neq 0$, then $r(a)$ is undefined and r has a vertical asymptote at $x = a$.

- If $p(a) = 0$ and $q(a) = 0$ and we can show that there is a finite number L such that

$$\lim_{x \to a} r(x) = L,$$

then $r(a)$ is not defined and r has a hole at the point (a, L).[1]

---

[1]It is possible for both $p(a) = 0$ and $q(a) = 0$ and for r to still have a vertical asymptote at $x = a$. We explore this possibility further in Exercise 5.5.4.9.

Activity 5.5.2. For each of the following rational functions, state the function's domain and determine the locations of all zeros, vertical asymptotes, and holes. Provide clear justification for your work by discussing the zeros of the numerator and denominator, as well as a table of values of the function near any point where you believe the function has a hole. In addition, state the value of the horizontal asymptote of the function or explain why the function has no such asymptote.

a. $f(x) = \dfrac{x^3 - 6x^2 + 5x}{x^2 - 1}$

b. $g(x) = \dfrac{11(x^2 + 1)(x - 7)}{23(x - 1)(x^2 + 4)}$

c. $h(x) = \dfrac{x^2 - 8x + 12}{x^2 - 3x - 18}$

d. $q(x) = \dfrac{(x - 2)(x^2 - 9)}{(x - 3)(x^2 + 4)}$

e. $r(x) = \dfrac{19(x - 2)(x - 3)^2(x + 1)}{17(x + 1)(x - 4)^2(x - 5)}$

f. $s(x) = \dfrac{1}{x^2 + 1}$

5.5.2 Sign charts and finding formulas for rational functions

Just like with polynomial functions, we can use sign charts to describe the behavior of rational functions. The only significant difference for their use in this context is that we not only must include all x-values where the rational function $r(x) = 0$, but also all x-values at which the function r is not defined. This is because it is possible for a rational function to change sign at a point that lies outside its domain, such as when the function has a vertical asymptote.

Example 5.5.7 Construct a sign chart for the function $q(x) = \frac{(x-2)(x^2-9)}{(x-3)(x-1)^2}$. Then, graph the function q and compare the graph and sign chart.

Solution. First, we fully factor q and identify the x-values that are not in its domain. Since $x^2 - 9 = (x - 3)(x + 3)$, we see that

$$q(x) = \frac{(x - 2)(x - 3)(x + 3)}{(x - 3)(x - 1)^2}.$$

From the denominator, we observe that q is not defined at $x = 3$ and $x = 1$ since those values make the factors $x - 3 = 0$ or $(x - 1)^2 = 0$. Thus, the domain of q is the set of all real numbers except $x = 1$ and $x = 3$. From the numerator, we see that both $x = 2$ and $x = -3$ are zeros of q since these values make the numerator zero while the denominator is nonzero. We expect that q will have a hole at $x = 3$ since this x-value is not in the domain and it makes both the numerator and denominator 0. Indeed, computing values of q for x near $x = 3$ suggests that

$$\lim_{x \to 3} q(x) = 1.5,$$

and thus q does not change sign at $x = 3$.

Thus, we have three different x-values to place on the sign chart: $x = -3$, $x = 1$, and $x = 2$. We now analyze the sign of each of the factors in $q(x) = \frac{(x-2)(x-3)(x+3)}{(x-3)(x-1)^2}$ on the various intervals. For $x < -3$, $(x - 2) < 0$, $(x - 3) < 0$, $(x + 3) < 0$, and $(x - 1)^2 > 0$. Thus, for $x < -3$, the sign of q is

$$\frac{---}{-+} = +$$

since there are an even number of negative terms in the quotient.

On the interval $-3 < x < 1$, $(x - 2) < 0$, $(x - 3) < 0$, $(x + 3) > 0$, and $(x - 1)^2 > 0$. Thus, for these x-values, the sign of q is

$$\frac{--+}{-+} = -.$$

Using similar reasoning, we can complete the sign chart shown in Figure 5.5.8. A plot of the function q, as seen in Figure 5.5.9, shows behavior that matches the sign function, as well as the need to manually identify the hole at $(3, 1.5)$, which is missed by the graphing software.

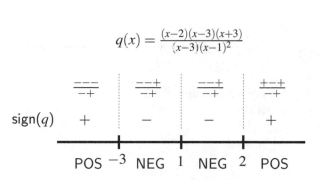

$$q(x) = \frac{(x-2)(x-3)(x+3)}{(x-3)(x-1)^2}$$

Figure 5.5.8: The sign chart for q.

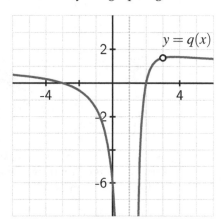

Figure 5.5.9: A plot of q.

In both the sign chart and the figure, we see that q changes sign at each of its zeros, $x = -3$ and $x = 2$, and that it does not change as it passes by its vertical asymptote at $x = 1$. The reason q doesn't change sign at the asympotote is because of the repeated factor of $(x - 1)^2$ which is always positive. □

To find a formula for a rational function with certain properties, we can reason in ways that are similar to our work with polynomials. Since the rational function must have a polynomial expression in both the numerator and denominator, by thinking about where the numerator and denominator must be zero, we can often generate a formula whose graph will satisfy the desired properties.

Activity 5.5.3. Find a formula for a rational function that meets the stated criteria as given by words, a sign chart, or graph. Write several sentences to justify why your formula matches the specifications. If no such rational function is possible, explain why.

a. A rational function r such that r has a vertical asymptote at $x = -2$, a zero at $x = 1$, a hole at $x = 5$, and a horizontal asymptote of $y = -3$.

b. A rational function u whose numerator has degree 3, denominator has degree 3, and that has exactly one vertical asymptote at $x = -4$ and a horizontal asymptote of $y = \frac{3}{7}$.

c. A rational function w whose formula generates a graph with all of the charac-teristics shown in Figure 5.5.10. Assume that $w(5) = 0$ but $w(x) > 0$ for all other x such that $x > 3$.

d. A rational function z whose formula satisfies the sign chart shown in Figure 5.5.11, and for which z has no horizontal asymptote and its only vertical asymptotes occur at the middle two values of x noted on the sign chart.

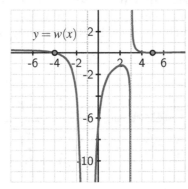

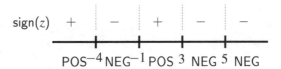

Figure 5.5.11: Sign chart for the rational function z.

Figure 5.5.10: Plot of the rational function w.

e. A rational function f that has exactly two holes, two vertical asymptotes, two zeros, and a horizontal asymptote.

5.5.3 Summary

- If a rational function $r(x) = \frac{p(x)}{q(x)}$ has the properties that $p(a) = 0$ and $q(a) = 0$ and

$$\lim_{x \to a} r(x) = L,$$

then r has a hole at the point (a, L). This behavior can occur when there is a matching factor of $(x - a)$ in both p and q.

- For a rational function $r(x) = \frac{p(x)}{q(x)}$, we determine where the function has zeros and where it has vertical asymptotes by considering where the numerator and denominator are 0. In particular, if $p(a) = 0$ and $q(a) \neq 0$, then $r(a) = 0$, so r has a zero at $x = a$. And if $q(a) = 0$ and $p(a) \neq 0$, then $r(a)$ is undefined and r has a vertical asymptote at $x = a$.

- By writing a rational function's numerator in factored form, we can generate a sign chart for the function that takes into account all of the zeros and vertical asymptotes of the function, which are the only points where the function can possibly change sign. By testing x-values in various intervals between zeros and/or vertical asymptotes, we can determine where the rational function is positive and where the function is negative.

5.5.4 Exercises

1. The graph below is a vertical and/or horizontal shift of $y = 1/x$ (assume no reflections or compression/expansions have been applied).

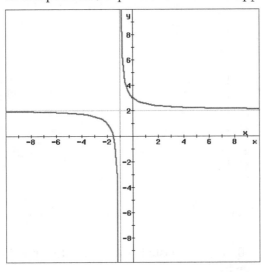

(a) The graph's equation can be written in the form

$$f(x) = \frac{1}{x + A} + B$$

for constants A and B. Based on the graph above, find the values for A and B.

(b) Now take your formula from part (a) and write it as the ratio of two linear polynomials of the form,

$$f(x) = \frac{Mx + C}{x + D}$$

for constants M, C, and D. What are the values of M, C, and D?

(c) Find the exact values of the coordinates of the x- and y-intercepts of the graph.

2. Find a possible formula for the function graphed below. The x-intercept is marked with a point located at $(1, 0)$, and the y-intercept is marked with a point located at $(0, -0.25)$. The asymptotes are $y = -1$ and $x = 4$. Give your formula as a reduced rational function.

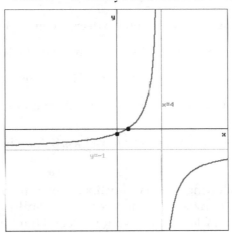

3. Find a possible formula for the function graphed below. The x-intercepts are marked with points located at $(5, 0)$ and $(-4, 0)$, while the y-intercept is marked with a point located at $\left(0, -\frac{5}{3}\right)$. The asymptotes are $y = -1$, $x = -3$, and $x = 4$. Give your formula as a reduced rational function.

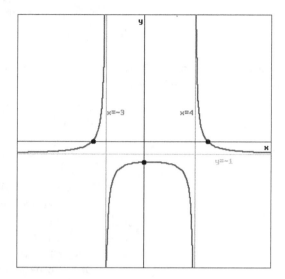

4. Let $f(x) = \dfrac{6x - 6}{7x + 4}$. Find and simplify $f^{-1}(x)$.

5. Let t be the time in weeks. At time $t = 0$, organic waste is dumped into a pond. The oxygen level in the pond at time t is given by

$$f(t) = \frac{t^2 - t + 1}{t^2 + 1}.$$

Assume $f(0) = 1$ is the normal level of oxygen.

(a) On a separate piece of paper, graph this function.

(b) What will happen to the oxygen level in the lake as time goes on?

☐ The oxygen level will continue to decrease in the long-run.

☐ The oxygen level will continue to increase in the long-run.

☐ The oxygen level will eventually return to its normal level in the long-run.

☐ It cannot be determined based on the given information.

(c) Approximately how many weeks must pass before the oxygen level returns to 80% of its normal level?

6. For each of the following rational functions, determine, with justification, the exact locations of all (a) horizontal asymptotes, (b) vertical asymptotes, (c) zeros, and (d) holes of the function. Clearly show your work and thinking.

a. $r(x) = \dfrac{-19(x + 11.3)^2(x - 15.1)(x - 17.3)}{41(x + 5.7)(x + 11.3)(x - 8.4)(x - 15.1)}$

b. $s(x) = \dfrac{-29(x^2 - 16)(x^2 + 99)(x - 53)}{101(x^2 - 4)(x - 13)^2(x + 104)}$

c. $u(x) = \dfrac{-71(x^2 - 13x + 36)(x - 58.4)(x + 78.2)}{83(x + 58.4)(x - 78.2)(x^2 - 12x + 27)}$

7.   Find a formula for a rational function that meets the stated criteria, with justification. If no such formula is possible, explain why.

   a. A rational function $r(x)$ in the form $r(x) = \frac{k}{x-a} + b$ so that r has a horizontal asymptote of $y = -\frac{3}{7}$, a vertical asymptote of $x = \frac{5}{2}$, and $r(0) = 4$.

   b. A rational function $s(x)$ that has no horizontal asymptote, has zeros at $x = -5$ and $x = 3$, has a single vertical asymptote at $x = -1$, and satisfies $\lim_{x \to \infty} s(x) = -\infty$ and $\lim_{x \to -\infty} s(x) = +\infty$.

   c. A rational function $u(x)$ that is positive for $x < -4$, negative for $-4 < x < -2$, negative for $-2 < x < 1$, positive for $1 < x < 5$, and negative for $x > 5$. The only zeros of u are located at $x = -4$ and $x = -2$. In addition, u has a hole at $x = 4$.

   d. A rational function $w(x)$ whose graph is shown in Figure 5.5.12.

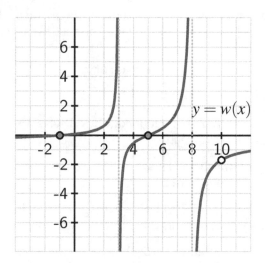

Figure 5.5.12: A plot of the rational function w.

8.   Graph each of the following rational functions and decide whether or not each function has an inverse function. If an inverse function exists, find its formula. In addition, state the domain and range of each function you consider (the original function as well as its inverse function, if the inverse function exists).

   a. $r(x) = -\dfrac{3}{x - 4} + 5$

   b. $s(x) = \dfrac{4 - 3x}{7x - 2}$

   c. $u(x) = \dfrac{2x - 1}{(x - 1)^2}$

   d. $w(x) = \dfrac{11}{(x + 4)^3} - 7$

9. For each of the following rational functions, identify the location of any potential hole in the graph. Then, create a table of function values for input values near where the hole should be located. Use your work to decide whether or not the graph indeed has a hole, with written justification.

   a. $r(x) = \dfrac{x^2 - 16}{x + 4}$

   b. $s(x) = \dfrac{(x - 2)^2(x + 3)}{x^2 - 5x - 6}$

   c. $u(x) = \dfrac{(x - 2)^3(x + 3)}{(x^2 - 5x - 6)(x - 7)}$

   d. $w(x) = \dfrac{x^2 + x - 6}{(x^2 + 5x + 6)(x + 3)}$

   e. True or false: given $r(x) = \frac{p(x)}{q(x)}$, if $p(a) = 0$ and $q(a) = 0$, then r has a hole at $x = a$.

10. In the questions that follow, we explore the average rate of change of power functions on the interval $[1, x]$. To begin, let $f(x) = x^2$ and let $A(x)$ be the average rate of change of f on $[1, x]$.

    a. Explain why A is a rational function of x.

    b. What is the domain of A?

    c. At the point where A is undefined, does A have a vertical asymptote or a hole? Justify your thinking clearly.

    d. What can you say about the average rate of change of f on $[1, x]$ as x gets closer and closer (but not equal) to 1?

    e. Now let $g(x) = x^3$ and $B(x)$ be the average rate of change of B on $[1, x]$. Respond to prompts (a) - (d) but this time for the function B instead of A.

    f. Finally, let $h(x) = x^3$ and $C(x)$ be the average rate of change of C on $[1, x]$. Respond to prompts (a) - (d) but this time for the function C instead of A.

Index

Colophon

This book was authored in PreTeXt.

CPSIA information can be obtained
at www.ICGtesting.com
Printed in the USA
LVHW061949111222
735016LV00004B/58